THE PLANT HUNTERS

THE ADVENTURES OF THE WORLD'S GREATEST BOTANICAL EXPLORERS

CAROLYN FRY

THE UNIVERSITY OF CHICAGO PRESS
Chicago and London

ONTENTS

RIGHT Stapelia campanulata *from Francis Masson's* Stapeliae novae, 1796.

INTRODUCTION

Nowadays, it's easy to fill our gardens with plants from around the world. Improved transportation and advanced propagation techniques mean that flowers that once only grew in the Amazon rainforest or Sahara desert are now available for anyone to buy at a garden center. Meanwhile, spices once confined to Pacific islands, fruits from tropical climes, and vegetables indigenous to South America have become commonplace in supermarkets the world over.

Plants were not always so easy to come by, however. The variety of species and cultivars we now take for granted is thanks to the exploits of numerous plant-collecting and horticultural pioneers. In the days before mass transport and commercialization, explorers risked their lives to obtain specimens from remote lands, nations fought wars over supplies, and empires expanded and contracted on the strength of trade in plant-based commodities.

The Egyptian pharaoh Hatshepsut set a precedent for moving plants around the planet when she transported incense trees from the land of Punt to Egypt in the fifteenth century B.C. At the time, fragrant tree resins such as myrrh and frankincense were sought as cures for various ailments and for use in religious ceremonies. Victorious Roman and Muslim armies later carried useful plants into the territories they conquered. Walnut trees and figs arrived in the U.K. with the invading Romans, for example.

The great seafaring expeditions of the fifteenth and sixteenth centuries were fueled by the desire to obtain spices. Christopher Columbus's crossing of the Atlantic, Vasco da Gama's journey to India, and Magellan's first circumnavigation of the Earth were all prompted by Western nations' desires for cloves, nutmegs, and mace. In time, as botanists learned to cultivate plants in new environments, obtaining seeds from far-away countries, smuggling them across borders, and setting up plantations became a way to make or break a nation's fortunes.

Scientists estimate that we have classified only three-quarters of plant species on the planet, so the role of the plant hunter is far from over. However, whereas we once moved plants around the world with little thought for the impacts of such actions, we are now more aware of the need to conserve habitats. With environmental destruction and climate change threatening as many as half the plant species with extinction, the onus is on today's botanists to preserve biodiversity. If they don't, we risk losing plants that could be useful to future generations before we have even discovered they exist.

This book tells the stories of plant hunters past and present whose botanical endeavors have helped shape the modern world. I have used a wide variety of sources to research the text, encompassing original diaries and journals along with secondary texts, such as biographies. Each chapter is richly illustrated with a mix of new photographs and historical illustrations. There are also beautifully reproduced facsimile copies of original documents, including some unpublished material from the archives of the Royal Botanic Gardens, Kew. These add another dimension to the stories of those pioneers whose explorations in remote and uncharted lands helped furnish us with the rich array of plants that we enjoy today.

CAROLYN FRY

ABOVE Trichonema speciosum *from* Curtis's Botanical Magazine, 1812.

BRINGING PLANTS FROM THE LAND OF PUNT

The first recorded plant hunter was Queen Hatshepsut, an Egyptian pharaoh who reigned for some 20 years in the fifteenth century B.C. She ruled as "King of Upper and Lower Egypt," the former being the land either side of the Nile Valley in the south, and the latter the river's delta in the north. Although recognized as female by her people, she adopted the image of a man by wearing a beard and is generally represented in works of art in male guise. Her reign was peaceful and prosperous, prompting a cultural renaissance that gave rise to celebrated paintings, sculptures, and temples. It is thanks to these that we know of Hatshepsut's plant-gathering exploits.

In the middle terrace of her temple at Deir el-Bahri, in Luxor's Valley of the Kings, are reliefs depicting an expedition of five ships sent by Hatshepsut to the land of Punt to gather exotic goods. The frieze shows sailors arriving in two boats and the chief of Punt, Parahu, stepping forward to greet the newcomers. There is a later scene illustrating the ships being loaded with ebony, apes, monkeys, dogs, panther skins plus "all goodly fragrant woods of God's Land, heaps of myrrh-resin, with fresh myrrh trees . . . " In all, 31 incense trees were brought back safely.

Scientists have long been divided over whether the mysterious land of Punt was south of Egypt in Africa or across the Red Sea in an Arabian country such as Yemen. They have used details illustrated in the relief at Deir el-Bahri together with a later depiction of Punt's inhabitants in the tomb of Ahmenmose (a chancellor of King Amenhotep III who lived around 80 years after Hatshepsut) to try and pinpoint Punt's location.

Much of the information presented is ambiguous. For example, the "Puntites" depicted on the Deir el-Bahri relief appear to have Arabian rather than African features, while those of Ahmenmose seem more African. Also, the presence of a giraffe on the Hatshepsut relief suggests an African location, while in the Ahmenmose tomb the image of a one-horned rhino indicates an Asian link and therefore an Arabian position. Meanwhile, frankincense, myrrh, and pistachio grow in both Africa and Arabia.

Scientists today are not certain that the resin was myrrh; it could alternatively have been frankincense or terebinth, a species of pistachio, which yields a sweet-scented gum from its bark.

Pinpointing Punt

Evidence from the tomb of the 17th-dynasty governor Sobeknakht provides the most compelling evidence to date for the location of Punt. When scientists investigated the tomb in 2003, they found it contained a 3,500-year-old inscription detailing an invasion of Egypt by the armies of Kush, which lay to the south. Among Kush's allies listed were the neighboring tribes of Wawat and Medjaw—plus Punt. It is therefore likely that Punt lay somewhere close to Kush, probably in the region of Eritrea, Ethiopia, or Somalia, rather than nearly 1,250 miles (2,000 kilometers) away in Arabia.

ABOVE *A statue of Queen Hatshepsut shows her in the guise of a man wearing a beard.*

BELOW *Djeser-Djeseru (meaning Holy of Holies), the mortuary temple of Queen Hatshepsut, is the most important part of Deir el-Bahri, which contains two other temples.*

ABOVE *A relief from Deir el-Bahri showing the loading of plants and other goods from Punt onto a boat bound for Egypt.*

Nowadays, we are more familiar with *Pistacia vera*, the species that produces the nuts we eat. However, it is known that fragrant resins were used extensively by the Egyptians. They burned them in temples, believing that gods descended through the atmosphere on the smoke. Similarly, they considered the souls of the dead would be transported on the vapors of incense burned in tombs. Some evidence suggests that terebinth was used as a varnish to protect tomb paintings. Whichever the species of trees shipped in from Punt, their transplant began a trend for shifting plants around the world that continues to this day.

LEFT *Almonds were a popular ingredient in Ancient Egyptian cosmetics.*

RIGHT *Frankincense is part of the genus* Boswellia*, and its resin is collected by cutting the bark and allowing it to bleed out and harden.*

Pistacia vera

from Henri-Louis Duhamel's Traité des arbres et arbustes . . . *, 1800. Pistacia vera produces pistachio nuts.*

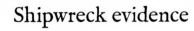

LEFT *These juniper berries were also found in Tutankhamun's tomb. Experts believe they may have been imported to Egypt from Greece.*

BELOW *Myrrh comes from the genus* Commiphora*, and its resin was used as an embalming ointment until the fifteenth century.*

BELOW *Dates were the most popular fruit in Ancient Egypt. These specimens were found in Tutankhamun's tomb.*

Shipwreck evidence

A shipwreck found at Uluburun, near Kas in southern Turkey, is proof that terebinth was an important commodity in ancient times. The ship—possibly a royal vessel—contained one tonne of terebinth resin in 150 Canaanite jars. Also on board were logs of Egyptian ebony (Dalbergia melanoxylon), storage jars containing pomegranates, plus elephant tusks and hippopotamus teeth. There were also two wooden writing boards, each comprising a pair of leaves joined with an ivory hinge. The main cargo was ten tonnes of cypriot copper. Scientists believe the ship sank around 1306 B.C.

Moving Plants from East to West

Between 10,000 and 3,500 years ago, several discrete human populations independently made the transition from a nomadic lifestyle to one based on agriculture. Changes to the climate, making wild food sources less reliable, are likely to have contributed to this switch in at least some locations. Ever since then, conquering nations have adopted plants they deemed useful as food, medicine, or building materials and used them to strengthen the resources of their empires. The Greek philosopher and scientist Theophrastus (372–287 B.C.) encouraged his contemporary Alexander the Great to send plants back from his warring campaigns across Persia, Egypt, Syria, Mesopotamia, Bactria, and the Punjab. When Theophrastus wrote *Enquiry into Plants*, the earliest surviving European treatise on botany, he included foreign plants, which may have been collected by botanists traveling with Alexander, alongside the local Mediterranean flora. One plant noted by Alexander's men in India, and now used globally, is cotton. They observed that the fabric woven out of fiber taken from the seeds was finer and whiter than any other cloth. As overland trade routes were set up to shift products such as cotton and silk garments, the movement of plants from East to West accelerated.

Origins of agriculture

The spread of modern humans, from their original home in Africa around 100,000 years ago, took around 50,000 years to encompass most of Eurasia. During this time, they were hunter-gatherers, relying on the plants and animals they encountered in the wild. The centers in which agriculture then developed, and the primary staple foods of each area, are: southwest Asia around the Fertile Crescent (barley and wheat); China (rice and millet); New Guinea (root and tree crops); sub-Saharan Africa (sorghum and pearl millet); Mesoamerica (maize and beans); eastern North America (seeded plants), and South America (potato and beans). Early farmers supplemented their diets with fruits and root crops, plus hunting and fishing. After A.D. 1500, European colonization accelerated the uptake of agricultural practices. Today, almost the entire human population depends on farmed food.

ABOVE *In this manuscript, which is held at the British Library, Aristotle instructs Alexander the Great to study plants and nature.*

LEFT *Triticum hybernum depicted by Pierre-Joseph Redouté in* La Botanique de Jean Jacques Rousseau, *1805. Emmer wheat and barley formed the basis for agriculture in the "Fertile Crescent."*

Ficus carica
from Antonio Targini Tozzetti's Raccolta di fiori, frutti ed agrumi, *1825. Figs were introduced to Britain by the Romans.*

RIGHT *Cotton fiber grows around the seed of the plant, which can be prone to disease and susceptible to pests.*

Later, as the Romans advanced they introduced new plants to lands they conquered. During the second century B.C., they began to take over formerly Greek lands, and by the end of the century held sway over the whole of the eastern Mediterranean. From 146 B.C., a great swathe of northern Africa, encompassing the modern countries of Morocco, Algeria, Tunisia, and Libya, came under Roman rule, and it was here that they grew great quantities of wheat, corn, barley, and olives to satisfy the appetites of their growing empire. They also planted vines in Lombardy, Tuscany, Syria, and Andalusia. Gaul (France and Belgium) and Germany west of the Rhine fell by 50 B.C., and Britain followed after A.D. 43. "The climate is unpleasant, with frequent rain and mist, but it does not suffer from extreme cold," wrote Publius Tacitus (c. A.D. 55-120) of the British climate. "The soil is fertile and is suitable for all crops except the vine, olive, and other plants requiring warmer climes." The list of plants introduced by the Romans to Britain includes the sweet chestnut (*Castanea sativa*), walnut (*Juglans regia*), fig (*Ficus carica*), leek (*Allium ampeloprasum* var. *porrum*), and opium poppy (*Papaver somniferum*).

Between the eighth and fifteenth centuries, Islamic conquest of much of the lands flanking the Mediterranean led to the widespread introduction of new crops from India and the Far East.

The globe-trotting potato

The ubiquitous potato is a widely traveled plant species that today grows in over 150 countries. The genus Solanum, *which includes the potato we now commonly eat (*Solanum tuberosum*), is native to South America. First cultivated in Peru and Bolivia from around 2,000 B.C., the potato was introduced to the Canary Islands around 1567 and to mainland Europe around 1570. The first known cultivation there took place in Seville between 1573 and 1576. Captain Cook took it to Australasia on his first circumnavigation of 1770. After cultivars of Chilean origin that suited cooler conditions were imported in the early nineteenth century, the potato spread rapidly across Europe and North America. The Irish became so reliant on the crop that when blight wiped out the harvests of 1845 and 1846, one million people died and another 1.5 million emigrated.*

Allium porrum

from François Regnault's
La botanique mise à la portée de tout
le monde, 1774. *The Romans considered
the leek superior to the onion and garlic.*

These included orchard fruits such as sour oranges, lemons, limes, pomelos, apricots, and bananas, plus rice, taro, and sugar cane. Some of these plants originally hailed from the Tropics, so raising them in relatively dry areas, such as Spain and Portugal, required expertise in creating irrigation systems. The Moors demonstrated their skills with water in showy flower gardens, growing oriental ornamentals such as tulips, yellow and white jasmine, narcissi, lilacs, and the Chinese rose. The most elaborate, with its fountains and water-lily pools, was the terraced garden of the Alhambra Palace complex in Spain. This was the last stronghold of the Muslim kings of Grenada before Spanish Christians forced the Moors to retreat south in 1492.

Jasminum nudiflorum

from Edwards'
Botanical Register, 1815-47.

IN SEARCH OF SPICES IN THE EAST

By the fifteenth century, Europeans had long been tempted by tales and products of the exotic East. Before the advent of ships capable of encircling the globe, short sea journeys and a tangle of overland routes were needed to bring culinary delights such as pepper, cloves, nutmeg, mace, ginger, and cinnamon from south and southeast Asia via a great swathe of Muslim-controlled land to Christian Europe. By the time these spices reached the markets of Venice, Bruges, and London, their price had been inflated by as much as 1,000 percent. Their great cost, far-away tropical source, their religious associations, and alleged medicinal powers—ranging from freshening the breath to making a small penis "splendid"—all rendered them highly desirable.

Towards the end of the thirteenth century, the Italian explorer Marco Polo resided for 17 years at the court of the Kublai Khan in China. After stopping in Indonesia on his return to Venice, he reported that Java had "nutmegs and cloves and other kinds of spices." Meanwhile, Nicolò de Conti's travels in the East during the first half of the fifteenth century were set down by Poggio Bracciolini. "Fifteen days' sail beyond [Java]," he noted, "two other islands are found: the one is called Sandai in which nutmegs and maces grow; the other is called Bandam; this is the only island in which cloves grow, which are exported hence to the Java islands." These were the Moluccas, lying to the east of Borneo. More sought-after spices, such as ginger, pepper, and cinnamon, could be found in India and Sri Lanka.

The Byzantine (or Eastern Roman) Empire came to an end when its capital, Constantinople, fell to the Turks in 1453. This led to a blockade of east-west overland trade routes to the mercantile capitals of southern Europe. Previously, Italians had bought Eastern products then sailed them to northern Europe via Portuguese ports. With this change in arrangements, the Portuguese lost previously guaranteed income, so they sought their own place in world trade. With the help of advances in maritime technology, the spice-hungry nation's sailors edged their way ever further down the west coast of Africa, hoping to find a route east. Meanwhile, the Genoese navigator Christopher Columbus approached the problem from a different direction. Given that the world was widely accepted to be round, he reasoned that sailing west across the Atlantic would provide an alternative route to the spice islands.

Columbus set off in August 1492, sponsored by Portugal's arch-enemy, Spain. When he reached the Caribbean in October, he thought he had found China. So blinkered was he that he even thought he saw spices where there were none. He described groves "all laden with fruit which the Admiral [Columbus] believed to be spices and nutmegs

Spices of life

Pepper (*Piper nigrum*) A perennial climbing vine, native to India's Malabar coast. The plant's spikes of peppercorns yield black, white, and green peppers, according to when they are picked and how they are prepared.

Nutmeg and mace (*Myristica fragrans*) A native of the Banda Islands of Indonesia, this tree yields two spices. Inside its apricot-like fruits, the glossy brown nutmegs are encased in a web of scarlet mace. Such was its value that fraudsters tried to sell fake nutmegs.

Sandalwood (*Santalum album*) Indigenous between eastern Java and Timor, this scented tree yields sandalwood essential oil. It is parasitic, which means it cannot survive unless it can draw nutrients from the roots of nearby trees.

Cinnamon (*Cinnamomum verum*) A small bay-like evergreen, native to southwest India and Sri Lanka. The spice is contained in the inner bark, which, once stripped from the tree, curls up into scrolls as it dries.

Clove (*Syzygium aromaticum*) Indigenous only to Ternate and Tidore islands in the Moluccas. This evergreen tree can grow to 40 feet (12.2 meters) and has glossy, aromatic leaves. The cloves grow in clusters and must be harvested before they overripen.

ABOVE *The Venetian writer and traveler Marco Polo noted that Java had nutmegs, cloves, and other spices.*

BELOW *A late-sixteenth-century map of the Spice Islands, now known as the Moluccas, engraved by Jodocus Hondius.*

MACHIAN   TIMOR alys MOTIR   POTTEBACKERS EYLANDT   Cleyn Mitere   MITERE   TERNATE   Hærn   Bay van Gilolo   Mauritius   Nassau

but they were not ripe and he did not recognize them." It was not until five years later, when Vasco da Gama rounded the Cape of Good Hope and sailed east to Calicot, the Indian epicenter of the spice-trading world, that the Portuguese were able to load up with the cargoes they longed for. However, they still sought the spice-bearing plants themselves, so in 1511, after taking over Malaysia's southern port of Malacca, they sent three ships in search of the Moluccas. Eventually, the expedition located the Banda islands, and filled two ships with nutmeg and mace. Francisco Serrão was despatched in the third ship to seek clove plants, which he eventually located on the northern Molucca island of Ternate.

After Columbus's encounter with the New World of South America, in 1494 the Treaty of Tordesillas divided newly discovered lands outside Europe between Spain and Portugal. Spain was entitled to anything lying west of the north-south meridian 370 leagues west of the Cape Verde islands, while Portugal was granted exploration

RIGHT *Christopher Columbus hoped to reach the Spice Islands by sailing west across the Atlantic; instead, he encountered the "New World" of South America.*

rights to anything east of the line. The trouble was, no one knew the Earth's circumference at this time, and no antimeridian was specified until a further treaty in 1529, so no one knew if the spice islands lay in Spanish or Portuguese territory. Thus it was that Spain sponsored Ferdinand Magellan to sail west around Cape Horn to try to reach the spice islands from the opposite direction.

The five-ship expedition left Spain in September 1519, with some 270 men. After seeking openings in the coast of South America that might provide a way through the continent, Magellan eventually sailed through the stormy southern straits that now carry his name and into relatively calmer seas that they named "Pacific." By now, only three ships remained. At this point, Magellan's men thought they were near the Moluccas, having no idea of the breadth of the ocean ahead. In fact, they were still some four months of sailing away. After Magellan died in a needless affray when they finally reached the western Pacific, his remaining men continued the search for the Moluccas. When they found the islands, much to the surprise of the Portuguese there, the Spaniards filled their hold with cloves and jubilantly sailed for home. Arguments continued to rage over which country had the right to the Moluccas' spicy bounty but eventually Spain capitulated. For the time being, at least, Portugal had won the spice race.

Botanical prize

The sailor who successfully led Magellan's men back to Spain after their leader's death was Juan Sebastián de Elcano. His reward was a coat of arms. On it, a globe set above two cinnamon sticks, 12 cloves, and three nutmegs is flanked by two Malay kings grasping branches of a spice tree, with the motto "Primus circumdedisti me," which translates as "You were the first to encompass me." The cloves brought back by the expedition paid for the entire trip, with a small profit remaining.

THE RISE OF THE PHYSIC GARDENS

The use of plants as medicines has a long history. The Indian text *Sushruta-Samhita* described 700 plants used as remedies as far back as 500 B.C., and the oldest Chinese herbal, the pharmacopoeia of Tzu-I, dates back to the same period. In the fifth century B.C., the Greek physician Hippocrates recorded Mediterranean plants with healing properties alongside exotics such as cinnamon, traded from India. Meanwhile, Dioscorides listed 650 species of plants around A.D. 50 in his *de materia medica libri quinque* ("five volumes concerning medical matter"). This work went on to form the cornerstone for many future pharmacopoeias.

By the fifteenth century, botanists had access to elementary classification systems and were becoming more knowledgeable about the properties of plants. Slowly, when it came to prescribing plants as remedies, contemporary observations began to take over from ancient teachings. The trade in medicines was unregulated; grocers, blood-letting surgeons, and apothecaries all dabbled in using plants to heal the sick. Distinguishing false from true "simples," the term given to a vegetable medicine made from a single constituent, was difficult to the untrained eye. This meant some unscrupulous traders substituted cheaper, ineffective ingredients for plants with true medicinal properties.

Francesco Bonafede, the first Reader in Simples at the University of Padua, Italy, campaigned for a *spetiaria* (pharmacy or spice shop) to act as a reference collection. He was granted his wish in 1545, when Padua opened a garden specifically to grow and categorize plants for use in science, research, and education. Considered the world's oldest

Rheum rhaponticum
from François Regnault's
La botanique mise à la portée de tout le monde, 1774.

botanic garden still in its original location, it is today recognized by UNESCO as a World Heritage Site. The University of Pisa had opened a similar garden in the preceding year, with plants grouped according to their properties such as scent and morphological characteristics, for example whether they exhibited bulbs or spines. Within decades, physic gardens had sprung up in Florence, Bologna, Leiden, Paris, and Oxford.

BELOW *A plan of Chelsea Physic Garden drawn in 1751 by John Haynes.*

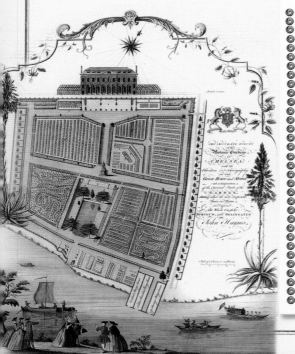

Nicholas Culpeper (1616-54)

Nicholas Culpeper is considered a figurehead of alternative medicine. Born in London, he returned there as an apprentice apothecary—or pharmacist—after studying at Cambridge. He did not become a "free man" licensee of the Apothecary Society, preferring to set up an independent business with an associate, Samuel Leadbetter. At the time, all prescriptions had to be prepared according to recipes set by the College of Physicians, but the pair questioned this convention and prepared their own, simpler medicines. In doing so, they upset other apothecaries and physicians who followed the guidelines. In 1649, Culpeper published A Physical Directory, *an English translation of the College of Physicians'* Pharmacopoeia, *an act that brought him more enemies. In 1653, he released* The English Physician *and* The Complete Herbal. *Both books sold well, the latter forming the cornerstone for herbalism in the English-speaking world. The Wakehurst Place estate, which is now managed by Kew Gardens, was once owned by a member of the Culpeper family.*

RIGHT *The Ancient Egyptians expelled worms with pomegranate.*

BELOW *Padua boasts the oldest botanical garden to have survived in its original location.*

Early Egyptian pharmacists

Most scientists believe scientific medicine began in the fifth century B.C. when the Greek physician Hippocrates introduced rational medicine based on diagnoses and a reasoned approach to treatment. However, in 2007, scientists deciphered 1,000 prescriptions written on four Egyptian papyri dating back to 1850 B.C. When cross-referencing the plants, animals, and minerals listed with modern medicinal directories, they found that 62 percent of the 284 ingredients were still in use in the 1970s. Remedies prescribed include: pomegranate or absinth to expel worms; coriander and cumin for flatulence; and celery seed to reduce swelling.

LEFT *Thomas Baskerville noted that the University of Oxford's Botanic Garden was popular with apothecaries treating the sick as well as healthy visitors.*

RIGHT *These coriander seeds date back to the time of Tutankhamun.*

By the middle of the sixteenth century, physicians had assumed control over other practitioners, and apothecaries in England were now able to dispense only remedies prescribed by licensed members of the College of Physicians. In London, the Worshipful Society of Apothecaries opened Chelsea Physic Garden in 1673 and wrangled with the physicians for the right to prescribe as well as dispense medicines. It finally gained this right in 1704. Apprentices, who included the famed herbalist Nicholas Culpeper, had to study for eight years before qualifying. They used the garden to help them learn how to recognize and prepare drug material correctly, and went on "herborizing" trips along the River Thames to gather new plants.

In time, physic gardens became natural repositories for foreign species brought back from the increasing number of expeditions to the far reaches of the planet. Garden plans of Padua drawn in 1571 and 1579 show that plants from the Balkan peninsula, the Levant, and America grew alongside natives from the Apennine Peninsula in its *horto rotondo* (circular garden). *Rheum rhaponticum* (rhubarb), *Cyclamen creticum*, and *Campanula saxatilis* were among the introductions. Meanwhile, when merchant apothecary John Watts was appointed to manage Chelsea Physic Garden in 1680, it was proposed he should plant "with foreign as well as local plants."

It was not long before the focus of botanical gardens shifted away from growing medicinal plants to the showcasing of exotics. Thomas Baskerville in his late-seventeenth-century *Account of Oxford Collectanea* noted that the University of Oxford's Botanic Garden was " . . . prouving serviceable not only to all Physitians, Apothecaryes, and those who are more immediately concerned in the practice of Physick, but to persons of all qualities serving to help ye diseased and for ye delight & pleasure of those of perfect health . . . " And Padua, having helped set the trend for medicinal gardens, began attracting tourists taking the Grand Tour. One visitor in 1786 was Johann Wolfgang von Goethe, who later wrote, "It is a pleasure and instructive to walk through vegetation that is strange to us."

Essentials oils prepared in the 1880s and now housed in the Economic Botany Collection at Kew Gardens in London are helping to authenticate the plant products being sold today. For example, pure samples of sandalwood from the collection have been used to test the provenance of oil being sold in India. Sandalwood is valued for its soothing and antibacterial properties on the skin but overexploitation is making the tree increasingly rare. Consequently, some traders peddle products that do not meet international standards, or they market synthetically produced oil as natural sandalwood. Kew's collection of more than 700 glass bottles of oils was among 10,000 plant-related *materia medica* (materials used for medicines) donated to Kew by the Royal Pharmaceutical Society of Great Britain in 1983.

Santalum album
(sandalwood) *from Joseph Jacob von Plenck's* Icones plantarum medicinalium, *1788-1812.*

LEFT Campanula saxatilis *from Hippolyte François Jaubert and Eduardo Spach's* Illustrationes plantarum orientalum, *1842-57.*

CAROLUS CLUSIUS AND TULIPOMANIA

When Carolus Clusius was born in France in 1526, European gardens had changed little since the eighth century. Most plants growing in his hometown of Arras would have been from northern Europe; he might have encountered a few southern European species, but plants of Asiatic origin would have been rarities. Hollyhocks, wallflowers, stocks, carnations, violets, cowslips, and marigolds were among the flowers that commonly colored gardens in his country. But there would have been no sunflowers, bulbous irises, or dahlias, and certainly no tulips.

Clusius studied in Belgium, Germany, and France before embarking on a plant-collecting trip to Spain and Portugal. He presented the results of his botanical explorations in a flora of the Iberian Peninsula. His botanical expertise gained him an invitation from the Habsburg Holy Roman Emperor Maximilian II to establish an imperial botanic garden in Vienna, in 1573, which he accepted. Having once been considered low-status activities practiced only by untrained apothecaries, plant collecting and botanical science were now fast becoming acceptable pastimes for the elite, and many new gardens were set up at this time.

As Europeans traveled into the Ottoman Empire, they encountered colorful plants they had not seen before and brought specimens back home. Ogier Ghiselin de Busbecq had been introduced to tulips while ambassador to the court of Suleiman the Magnificent, a role he was appointed to by Maximilian II's predecessor, Ferdinand I. He supplied some seeds and bulbs of tulips for the Vienna garden. Clusius began experimenting with growing them and, in time, became an expert. He wrote up his findings in the first-ever treatise on the

tulip, *Rariorum Plantarum Historia*, published in 1601. In it, he classified tulips as early, late, or intermediate flowering, and described their colors, shape, and other features. "The color of the reds is usually rather deep and almost blackish, but sometimes of a lighter shade and very elegant," he wrote of one.

Tulips had long enthralled inhabitants of the lands in which they grew wild, namely Turkey, Persia, the Crimea, the Caucasus, the Levant, Chitral, Afghanistan, and the Siberian steppes. Tiles and other ornaments portraying tulips decorated Anatolian Seljuk buildings as far back as the twelfth century. In the thirteenth century, the Persian poet Sa'adi had described a visionary garden as: "The murmur of a cool stream/bird song, ripe fruit in plenty/bright multicolored tulips and fragrant roses . . . "

What made the flowers so magical was their sheer range of shapes and colors, and the seemingly random way in which they changed from plain blooms to multicolored ones. Clusius observed this in some of the plants he grew. Tulips exhibiting such color changes were described as "broken." The effect is now known to be caused by a virus spread by aphids, but at the time it was considered a wonder of nature. As a result, broken tulips became highly sought-after.

RIGHT *As tulips became more popular, vases were designed specially to display prized blooms.*

Tulipa gesneriana dracontia and *T. g. variegata* from Antonio Targini Tozzetti's Raccolta di fiori, frutti ed agrumi, 1825.

Tulipomania's legacy

There are now known to be around 120 species of tulip, mostly native to Central Asia. They belong to the family Liliaceae, *genus* Tulipa. *There are some 2,300 extant named cultivars of garden tulip. Since 1996, tulips have been classified into 13 divisions: Single Early, Double Early, Triumph, Darwin Hybrid, Single Late, Lily-flowered, Fringed, Viridiflora, Rembrandt, Parrot, Double Late, Kaufmanniana, Fosteriana, Greigii, and Species. Names are officially registered in the* Classified List and International Register of Tulip Names, *published in The Netherlands by the Royal General Association of Bulbgrowers.*

Clusius later worked in Frankfurt and then headed for Leiden in The Netherlands to lay out a physic garden at the city's new university. He took with him the best collection of tulips in western Europe, and extended their reach by sending bulbs to acquaintances. "I owe you undying gratitude for the outstanding benevolence you bestowed on me when I was in Leiden last year;

L. Tulipa alba cum rubr . . .

drawn by Crispijn van de Passe the Younger for Hortus Floridus (1614).

for your services and your merits, and, above all, for the precious bulbs," wrote Bergen Høyer in 1597. People soon realized that they could profit from growing tulips, and after Clusius's most valuable tulips were stolen from his garden, the flowers spread across The Netherlands.

Entrepreneurs began establishing tulip nurseries in Holland at the start of the seventeenth century. They initially sold bulbs in large quantities to estate owners, but by the 1620s, single-colored varieties were selling for 12 florins a pound (at a time when the average annual income was about 150 florins). Soon, tulip collecting took off and in 1624, "Semper augustus" bulbs were fetching 1,200 florins each. Within a year the price had more than doubled and the market was invaded by speculators keen to cash in on this "Tulipomania." The highest price ever *asked* for "Semper augustus" was quoted in *Nederlandsch Magazijn* as 13,000 florins for a single bulb, in those days more than the cost of an expensive house overlooking the canals of central Amsterdam.

By 1637, the bubble had reached bursting point and as sellers began to outnumber buyers, the market collapsed. For a time, tulips became worthless, even despised. The Professor of Botany at Leiden, where Clusius had first planted tulips in Dutch soil, came to hate the flowers so much that he attacked them with his cane. By now Clusius was dead, but his passion for tulips had changed the appearance of northern European gardens forever and sowed the seeds for the bulb industry that thrives in The Netherlands to this day. Over 49,000 acres (20,000 hectares) are now planted with flower bulbs each year, and, in 2007, the turnover from *Tulipa* sold at auctions exceeded 260 million dollars.

ABOVE *These tulips were painted in watercolors by Simon Verelst (1644–c. 1721) and are part of the Sir Arthur Church collection at the Royal Botanic Gardens, Kew.*

Carolus Clusius (1526–1609)

One of the reasons we know so much about Carolus Clusius's work is because of his prolific letter-writing. Over the course of his life, he corresponded with a network of some 300 different people across Europe. The recipients of his letters included: aristocratic collectors and rich patrons, such as Charles de Saint Omer in the southern Netherlands, Lord Zouche in England, and Ludwig VI, Elector of the Palatinate in Germany; fellow botanists or physicians, such as Joachim Camerarius, Felix Platter, and Ulisse Aldrovandi; diplomats, such as Ogier Ghiselin de Busbecq; and apothecaries, such as Jean Mouton. The information and specimens Clusius received from his correspondents enabled him to describe more exotic species than any other sixteenth-century botanist. In 2004, the Scalinger Institute and Leiden University Library digitized 1,300 letters he received and made them freely available online.

CLUSIUS

THE TRADESCANTS MAKE PLANT HUNTING A CAREER

By the late sixteenth and early seventeenth centuries, planting gardens was in the West becoming an indulgent hobby for wealthy gentlemen. As developing trade links brought news of the diversity of botanical riches that existed in foreign parts, the owners of large estates vied to create the most unusual and exotic collections of plants. One such gentleman was Robert Cecil, the First Earl of Salisbury, who began developing the garden at Hatfield House in Hertfordshire, England, in 1610. He employed John Tradescant as gardener, sending him to The Netherlands, Belgium, and France to obtain tulip bulbs, rose bushes, and cherry, pear, quince, mulberry, and orange trees. In doing so, he helped elevate the status of plant hunting from an enjoyable pastime to a lucrative profession.

Tradescant went on several collecting missions for Cecil and subsequent employers. From Russia, he gathered roses "single and

FAR LEFT *John Tradescant the Elder, who gathered plants and curiosities from Europe and Russia.*

LEFT *John Tradescant the Younger introduced a number of species from America, such as magnolia, phlox, and aster.*

Ark" were crocodile eggs, a lion's head, a banana, sandals of wood from China, a Turkish toothbrush, and a "Booke of Mr Tradescant's choisest Flowers and Plants, exquisitely limned in vellum." The first museum in England to open to the public, the Ark was a place where "a Man might in one daye behold and collect into one place more Curiosities then hee should see if hee spent all his life in Travell." Outside, the garden was a showcase for clematis, hyacinths, oleanders, and other botanical treasures from afar.

Tradescant died in 1638, leaving his son to inherit the property in Lambeth along with his position at Oatlands. When the death was announced, John Tradescant the Younger was already following in his father's plant-hunting footsteps on the first of three journeys to Virginia "to gather up all raritye of flowers, plants, shells &c." Virginia

RIGHT *The palace of Oatlands was the home of King Charles I. The site is now occupied by a hotel.*

muche like oure sinoment [cinnamon] rose;" on a journey aimed at destroying the fleet of the troublesome Barbary Pirates of Algiers, he picked up a "rough starry headed trefoil," and while traveling to France as an engineer for the Duke of Buckingham, he brought back the "Broad Leafed Sea Wormwood." As well as collecting plants, he brought back shells, fossils, animals, and unusual trinkets. Such was his success as a traveler, gardener, and collector of curiosities that by 1632 he was employed by King Charles I and Queen Henrietta Maria as Keeper of the Gardens, Vines, and Silkworms at Oatlands Palace. Built by Henry VIII, Oatlands then stood beside the River Thames between Walton and Weybridge in Surrey.

Tradescant acquired a house in Lambeth, south London, in which he displayed his burgeoning collection of souvenirs. Among the animals, plants, fish, and fossils on display in "Tradescant's

Tradescantia virginiana

from Pierre-Joseph Redouté's Les liliacées, *1802-16. It was first listed in the elder Tradescant's collection in 1629.*

The myth of the vegetable lamb

One object in the Museum of Garden History is shaped like a miniature sheep. It is the fabled "vegetable lamb" or "Barometz" of Tartary. The myth of a living plant that produced lambs as its fruit was first recorded in the fifth century and persisted until the seventeenth century. People believed it was an animal rooted by a flexible stem to the Earth. In 1605, Claude Duret of Moulins devoted a chapter to the "Barometz of Scythia or Tartary" in his Histoire Admirable des Plantes; *in 1629, the English king's herbalist, John Parkinson, illustrated it in his book* Paradisi in Sole Paradisus Terrestris; *and the Tradescants claimed in 1656 to have "a very small part" of Barometz skin. The "lamb" was in fact the hairy root of the Asian fern* Cibotium barometz, *sculpted to look like a sheep.*

had been founded in 1607 as the first permanent English colony in the New World, and plants from the area were soon sprouting from English soil. Tradescant introduced the tulip tree and a yucca plant to the United Kingdom and also supplemented the Ark with artifacts from America, including Powhatan's mantle, a shell-decorated animal skin garment made by the Virginia Indians. After Tradescant the Younger died, Elias Ashmole acquired the collection and presented it to the University of Oxford in 1683. Some pieces, including Powhatan's mantle, still reside in the university's Ashmolean Museum.

Rosa acicularis
from John Lindley's Rosarium monographia, 1820. *It was collected by the elder Tradescant in Russia.*

A world of wonders in one closet shut,
These famous antiquarians that had been
Both Gardiners to the Rose and Lily Queen,
Transplanted now themselves, sleep here & when
Angels shall with their trumpets waken men,
And fire shall purge the world, these three shall rise
And change this garden then for Paradise.

Both John Tradescants were buried in a family tomb at St. Mary-at-Lambeth Church, which still stands in London beside the Thames opposite the Houses of Parliament. The church was deconsecrated in 1972 and would have been demolished had not John and Rosemary Nicholson rediscovered the Tradescant grave and founded the Museum of Garden History in memory of the father and son botanists. The corner of the graveyard in which the tomb now stands is planted with species that grew in the Tradescant garden in 1656. The grave itself is decorated with a tree at each corner, a crocodile, shells, plus columns and pyramids swept up in a maelstrom.

The epitaph reads:

Know stranger, ere thou pass, beneath this stone
Lye John Tradescant, grandsire, father, son,
The last dy'd in his spring, the other two
Liv'd till they had travell'd Orb and nature through
As by their choice Collections may appear,
Of what is rare, in land, in sea, in air,
Whilst they (as Homer's Illiad in a nut)

Fritillaria imperialis
from Pierre-Joseph Redouté's Les liliacées, 1802-16.

LEFT *The tomb of the Tradescants can now be seen by visitors to the Museum of Garden History in Lambeth on the south bank of the River Thames in London.*

SHOWCASING EXOTIC PLANTS IN EUROPE

In the eighteenth century, botanical gardens began to take on a broader scope than the early medicinal plots. As exploration revealed ever more plant species previously unknown to the West, European nations raced to discover new lucrative botanical commodities.

France's foremost botanical garden is the Jardin des Plantes in Paris. Founded in 1626 as the Jardin du Roi ("Garden of the King"), it was initially a medicinal garden. As France extended its overseas territories, the garden worked with the navy to provide medicinal plants required on long sea journeys. Its collections swelled as plants were brought to the garden from foreign ports of call. In 1636, it was growing 1,800 species; by 1665 the number reached 4,000.

When France lost its position as a major colonial power after the Seven Years' War, Louis XVI's marine ministry commissioned Jean-François de Galaup, Comte de La Pérouse, to sail around the world undertaking scientific studies. The King's chief gardener, André Thoüin, gave instructions for sowing European plants in new places and climates, and for returning newly discovered species to the royal garden in Paris. In 1755, Spain's King Ferdinand VI ordered the creation of El Real Jardín Botánico del Soto de Migas Calientes, forerunner of today's Real Jardín Botánico de Madrid. The garden was moved to its present location, the Paseo del Prado (meaning the "Walk by the Meadow"), in 1781, during the reign of Ferdinand's successor, Charles III. Like the medicinal gardens of Renaissance Italy, the garden was designed to be educational: a place where scholars could learn about botany. But from the outset, it also sought to promote expeditions for the discovery of new plant species and to classify them when they arrived in Spain from remote shores.

In the 1570s, Spain had sent physician Francisco Hernández to Mexico and other Spanish possessions in the New World to gather information on the local uses of plants. This is now recognized as the first expedition of natural history ever sent out by a government. Two hundred years later, the country was keen to invest in more scientific forays. During the garden's early years, it dispatched José Ortega around the Iberian Peninsula to collect plants. Then, in the last quarter of the eighteenth century, it sent Jose Antonio Pavón Jiménez and Hipólito Ruiz López to Peru and Chile; sponsored the priest and naturalist José Celestino Bruno Mutis y Bosío in his botanical studies of New Grenada (now Columbia); sent the Royal Scientific Expedition to New Spain (now Mexico) to follow up on Hernández's work; and launched the round-the-world Malaspina Expedition. During Antonio José Cavaniles's directorship of the garden in the first years of the nineteenth century, more than 150 plants from the New World were growing there.

The seeds of the United Kingdom's Royal Botanic Gardens at Kew were sown in the same decade as the Madrid garden. At the time, two royal estates lay side by side beside the River Thames in west London. King George II inhabited Richmond, the plot that lay closer to the river, while his son, Frederick, Prince of Wales, lived on the neighboring estate of Kew. Both royal households planted trees and landscaped their estates, as was fashionable for wealthy landowners of the time. When Frederick died in 1751, his wife, Augusta, continued to develop the 100 acres (40 hectares) she had inherited with the help of Lord Bute, an aristocrat skilled in gardening. Then, in 1759, she decided to create a physic garden. Her aim, like King Ferdinand VI's,

LEFT *The roots of Spain's Real Jardín Botánico in Madrid date back to 1755.*

ABOVE *A packet of cinchona seeds sent back to Europe by Jose Antonio Pavón Jiménez. His signature can just be seen on the left.*

LEFT *An engraving showing the pagoda at Kew that appeared in William Chambers's* Plans, elevations, sections, and perspective views of the gardens and buildings at Kew in Surry [*sic*], 1763.

Echinocactus cornigerus

from Revue des Cactées, 1829.

was to showcase exotics. The garden should, she declared ambitiously, "contain all the plants known on Earth."

The oldest known trees at Kew Gardens are called the "Old Lions." Planted around 1762, within a few years of Princess Augusta founding the garden, they comprise *Ginkgo biloba* (maidenhair tree), *Styphnolobium japonicum* (pagoda tree), *Platanus orientalis* (oriental plane), *Robinia pseudoacacia* (false acacia), and *Zelkova carpinifolia* (Caucasian elm). Some came from a neighboring estate at Whitton that belonged to Lord Bute.

Once established, both Kew and Madrid went on to set up supporting gardens in areas under their country's rule. Spain set up offshoots in Mexico, Valencia, and on the island of Tenerife, to experiment with growing plants from the New World in climates more suitable than that of Madrid. Kew eventually worked with many offshoot gardens, including gardens in India, Ceylon (now Sri Lanka), Trinidad, St. Vincent, and Jamaica. Both gardens also benefited from the exchange of plants between the best botanists of the day. Madrid received green parcels from Dr. Fothergill of London, from botanic gardens in France and Italy, and from the colonies of Peru, Cuba, and Mexico. Meanwhile, Kew received botanical offerings from the amateur naturalist John Ellis and botanist Peter Collinson, among others. Such was the influx of plants to west London that Lord Bute bragged to the Governor of Georgia that "the Exotic Garden at Kew is by far the richest in Europe . . . getting plants and seeds from every corner of the habitable world."

BELOW *The pagoda tree (*Styphnolobium japonicum*) at Kew Gardens dates back to 1762, a few years after the garden was founded.*

BELOW RIGHT *A nineteenth-century depiction of the Jardin des Plantes in Paris.*

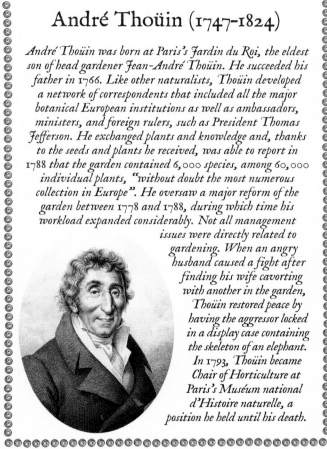

André Thoüin (1747–1824)

André Thoüin was born at Paris's Jardin du Roi, the eldest son of head gardener Jean-André Thoüin. He succeeded his father in 1766. Like other naturalists, Thoüin developed a network of correspondents that included all the major botanical European institutions as well as ambassadors, ministers, and foreign rulers, such as President Thomas Jefferson. He exchanged plants and knowledge and, thanks to the seeds and plants he received, was able to report in 1788 that the garden contained 6,000 species, among 60,000 individual plants, "without doubt the most numerous collection in Europe". He oversaw a major reform of the garden between 1778 and 1788, during which time his workload expanded considerably. Not all management issues were directly related to gardening. When an angry husband caused a fight after finding his wife cavorting with another in the garden, Thoüin restored peace by having the aggressor locked in a display case containing the skeleton of an elephant. In 1793, Thoüin became Chair of Horticulture at Paris's Muséum national d'Histoire naturelle, a position he held until his death.

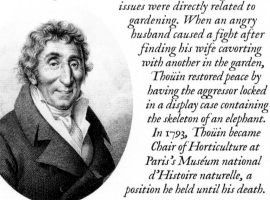

CARL LINNAEUS AND THE NAMING OF PLANTS

The name of every plant and animal species comprises two words in Latin. The sunflower is *Helianthus annuus* L.; the white oak is *Quercus alba* L.; and the coconut palm is *Cocos nucifera* L. The first word denotes the genus it belongs to, the second the species. The genus and the species are followed by the author(s) of the plant name: in these cases, the eighteenth-century Swedish botanist Carl Linnaeus, abbreviated to "L." Passionate about botany from a young age, he dedicated his life to classifying plants and animals and was the first to make consistent use of the binomial system for naming species that is still in place today.

Linnaeus drew on the work of several seventeenth-century scientists to create a classification system incorporating the categories of kingdom, class, order, genus, and species. He based his system on the characteristics of a plant's sexual organs. He identified 24 classes according to the number, length, and distinctive features of pollen-bearing stamens (male parts) and divided these into orders according to their pistils (female parts). He used his system to compile a comprehensive list of all plants known at that time worldwide and published it in 1753 in a volume called *Species Plantarum*. This is now considered the starting point for the modern system of botanical nomenclature; names used prior to the publication are not valid.

Linnaeus was criticized by some of his contemporaries for choosing to base his classification on sexual organs. One opponent, botanist Johann Siegesbeck, called it "loathsome harlotry," but Linnaeus got his revenge when he named a weed *Siegesbeckia orientalis* L. after him. Linnaeus's scheme was quickly adopted, as the existing method of naming plants purely on descriptions was becoming unworkable with the multitude of new species being discovered at the time. For example, the tomato, which Linnaeus named *Solanum lycopersicum*, had formerly been known as *Solanum caule inermi herbaceo, foliis pinnatis incisis, racemis simplicibus*—the solanum with the smooth stem which is herbaceous and has incised pinnate leaves.

Although the basis of Linnaeus's classification system is still used, the understanding of genealogical relationships between different plants has changed considerably over the centuries. The publication of Charles Darwin's *On the Origin of Species* in 1859 caused the first major upheaval by demonstrating that life forms had evolved over time through the process of natural selection. Today, modern techniques of DNA-sequencing are complicating the botanical family tree further: roses that were once linked to saxifrages are now considered closer to nettles, while papayas formerly linked to passion flowers are deemed more akin to cabbages. Some scientists have called for the Linnaean system of naming plants to be replaced by an International Code of Phylogenetic Nomenclature, or "PhyloCode." This would organize and name plants on the basis of their evolutionary relationships.

Scientists have now identified 400,000 plant species but believe there may be several million more still left to be

RIGHT *The passion flower* (Passiflora) *is a genus of about 500 species, and its name refers to the passion of Christ on the cross, with various parts of the flower representing elements of the crucifixion.*

BELOW Helianthus annuus L., *the sunflower, is one of many plants named by Linnaeus.*

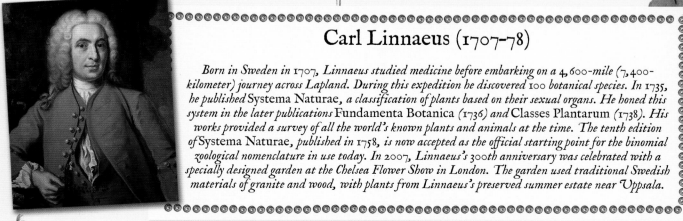

Carl Linnaeus (1707-78)

Born in Sweden in 1707, Linnaeus studied medicine before embarking on a 4,600-mile (7,400-kilometer) journey across Lapland. During this expedition he discovered 100 botanical species. In 1735, he published Systema Naturae, *a classification of plants based on their sexual organs. He honed this system in the later publications* Fundamenta Botanica *(1736) and* Classes Plantarum *(1738). His works provided a survey of all the world's known plants and animals at the time. The tenth edition of* Systema Naturae, *published in 1758, is now accepted as the official starting point for the binomial zoological nomenclature in use today. In 2007, Linnaeus's 300th anniversary was celebrated with a specially designed garden at the Chelsea Flower Show in London. The garden used traditional Swedish materials of granite and wood, with plants from Linnaeus's preserved summer estate near Uppsala.*

discovered. In 2008, researchers from Kew Gardens and Imperial College, London, announced they had identified a "barcode" gene, enabling them to distinguish between the majority of plant species on Earth. While this should simplify the task of cataloging plants in species-rich areas such as rainforests, the challenge will be to do so before climate change and environmental degradation render them extinct. The plant *Linnaea borealis*, named after Linnaeus, is a case in point. Its populations in the U.K. have declined so much, due to the loss of pine woodland habitat, that its survival in the country is now in doubt.

Rosa virginiana
from François Plée's Types de chacque famille et des principaux genres des plantes . . . , 1844-64.

LEFT *The title page of Charles Darwin's groundbreaking book* On the Origin of Species.

BELOW *The family home of Carl Linnaeus, called Linnegard, which Carl's father, Nils, used in Latin form as his surname.*

Saxifraga granulata
from François Plée's Types de chacque famille et des principaux genres des plantes . . . , 1844-64.

Sir Joseph Banks

Joseph Banks was a philanthropist interested in natural history, science, and global affairs. His love of nature grew out of exploring the countryside around his family's ancestral home in Lincolnshire. By the age of 17, he already had a herbarium (a collection of dried plant specimens), and went to Oxford University hoping to hone his skills as a botanist. On inheriting considerable wealth from his father in 1761, aged 18, he quit university without a degree and two years later joined HMS *Niger* on a voyage to collect plants, rocks, and animals from Newfoundland and Labrador. In 1768, having already been elected as a Fellow of the Royal Society due to his potential to excel in the natural sciences, he was invited to join Captain James Cook on an expedition to the South Pacific.

The main aim of the voyage was to observe the transit of the planet Venus and calculate the size of the solar system. However, the mission that Banks and his eight staff planned to fulfill was to record the natural history of the places the *Endeavour* visited. These included South America, Tahiti, New Zealand, Australia, and Java. When the expedition returned in 1771, Banks brought home 3,600 dried specimens of plants, of which 1,400 were new to science. There were 14 plants introduced to cultivation in Britain in 1771, including *Haloragis erecta* and *Leptospermum scoparium* from New Zealand, and *Eucalyptus gummifera* and *Dianella caerulea* from Australia. Over 50 plants were named subsequently from herbarium specimens. These include *Astelia banksii* (from New Zealand), *Callistemon viminalis*, and *Melaleuca nodosa* (from Australia). Banks planned to join a second voyage departing the following year but his request for 15 staff, including two French horn players, met with short shrift from Cook, who felt the modifications needed to accommodate them would make the ship top heavy. Banks then withdrew from the venture.

Banks thereafter remained mostly in Britain. However, he was responsible for transferring numerous

Pittosporum fulvum
by E. D. Smith from Robert Sweet's
Flora Australasica, 1827–8.

plants to and from the country through his role as unofficial director of Kew Gardens from around 1773. He sent Kew botanist Francis Masson to gather plants from the Cape of Good Hope in 1772, Archibald Menzies to the northwest coast of America in 1791, and William Kerr to China in 1803. He also despatched Allan Cunningham and James Bowie to South America in 1814, from

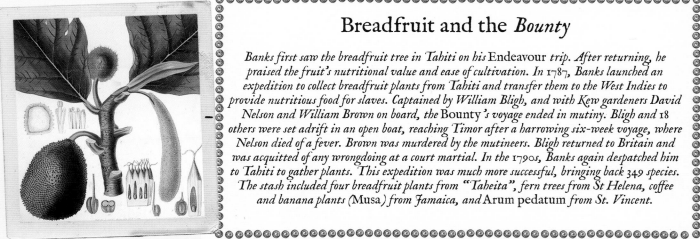

Breadfruit and the *Bounty*

Banks first saw the breadfruit tree in Tahiti on his Endeavour *trip. After returning, he praised the fruit's nutritional value and ease of cultivation. In 1787, Banks launched an expedition to collect breadfruit plants from Tahiti and transfer them to the West Indies to provide nutritious food for slaves. Captained by William Bligh, and with Kew gardeners David Nelson and William Brown on board, the* Bounty's *voyage ended in mutiny. Bligh and 18 others were set adrift in an open boat, reaching Timor after a harrowing six-week voyage, where Nelson died of a fever. Brown was murdered by the mutineers. Bligh returned to Britain and was acquitted of any wrongdoing at a court martial. In the 1790s, Banks again despatched him to Tahiti to gather plants. This expedition was much more successful, bringing back 349 species. The stash included four breadfruit plants from "Taheita", fern trees from St Helena, coffee and banana plants (*Musa*) *from Jamaica, and* Arum pedatum *from St. Vincent.*

where they journeyed to Botany Bay, Australia, and the Cape of Good Hope, South Africa, respectively. Masson furnished our gardens with geraniums and the bird-of-paradise flower *Strelitzia reginae*; Kerr introduced *Kerria japonica*; and Cunningham brought back many species of *Eucalyptus*, *Acacia*, and *Pittosporum*.

Banks gave his collectors strict instructions. For example, in a letter he wrote to Archibald Menzies in 1791, he demanded:

"When you meet with curious or valuable plants which you do not think likely to be propagated from seeds in His Majesty's Garden, you are to dig up proper specimens of them, plant them in the glass Frame provided for that purpose, and use your utmost endeavours to preserve them alive till your return, and you are to consider every one of them, as well as all Seeds of Plants which you shall collect during the voyage, as wholly and entirely the property of His Majesty, and on no account whatever to part with any of them, or any cuttings, Slips, or parts of them, for any purpose whatever but for His Majesty's use."

As well as introducing plants to the United Kingdom, Banks heavily influenced the choice of plants taken with the first settlers to Australia. Having experienced the land for himself, he said he "had no doubt that the Soil of many Parts of the Eastern Coast of New South Wales between the Latitudes of 30 & 40 is sufficiently fertile . . ." He equated the climate there to that of Toulouse in France and put together a "*portmanteau biota*" of suitable European vegetables, herbs, berries, fruits, and grains for the settlers to grow. It was this European flora that ultimately helped the newcomers thrive. In his book *The Settlement at Port Jackson*, British Marine Watkin Tench observed: "Vines of every sort seem to flourish: melons, cucumbers and pumpkins run with unbounded luxuriancy; and I am convinced that the grapes of New South Wales will, in a few years, equal those of any other country . . ."

Banks's vision of plants playing a central role in shaping societies was also put into practice through the botanical gardens set up in Britain's colonies. He regularly exchanged seeds and advice with the directors, often Kew-trained, of gardens in India, Ceylon (now Sri Lanka), St. Vincent, Trinidad, and Jamaica. The French naturalist Baron Georges Cuvier remarked of him, "He has spread over all gardens of Europe the seeds from the Southern Sea as he has distributed ours in the Southern Sea." Made a baronet in 1781, Banks helped found the Royal Horticultural Society in 1804 and was President of the Royal Society for four decades. Today, his collections of plants, insects, and shells reside at the Natural History Museum in London, and his name lives on in the volcanic Banks Islands of the Pacific and the genus *Banksia*, the Australian honeysuckle.

RIGHT Vitis vinifera *from Antonio Targini Tozzetti's* Raccolta di fiori, frutti ed agrumi, *1825.*

Strelitzia reginae
the bird-of-paradise flower painted by
Francis Bauer (1758–1840) from the archives
of the Royal Botanic Gardens, Kew.

Banks's Florilegium

This collection of botanical line engravings records plants gathered by Sir Joseph Banks and the Swedish botanist Daniel Solander while on Cook's first voyage around the world. They were created from watercolors painted by Sidney Parkinson, one of the expedition's two official artists. When Parkinson died during the voyage in 1771, he had only completed 238 watercolors, but Banks later employed 18 engravers to produce more than 700 plates from Parkinson's drawings and dried specimens. He later bequeathed the copperplates used to make the florilegium to the British Museum in London. A full edition of the engravings was published for the first time in 1989.

GATHERING PLANTS FROM SOUTH AMERICA

After Christopher Columbus made landfall in South America at the end of the fifteenth century, European countries were eager, according to the Spanish botanist Nicolás Monardes, to exploit the "diuerse and sundrie hearbes, trees, oyles, plantes, and stones" of this "Newe Founde Worlde." However, the Treaty of Tordesillas divided these lands solely between Portugal and Spain. Only when France requested to send an expedition to Peru, to measure the length of a degree of the meridian close to the Equator, did Spain allow outsiders in. A team led by Charles Marie de la Condamine set off in 1735 and spent eight years working to ascertain the exact shape of the Earth. At the end of the expedition, de la Condamine traveled along the Amazon, his journal entries providing outsiders with their first glimpse of what the South American interior was like. "At Borja I found myself in a new world, separated from all human intercourse, on a fresh-water sea, surrounded by a maze of lakes, rivers, and canals, and penetrating in every direction the gloom of an immense forest . . . New plants, new animals, and new races of men were exhibited to my view."

One tree that intrigued him was the "heavea," which oozed waterproof latex, called caoutchouc, from its bark. He noted that "when fresh it could, by means of molds, take any shape given to it, at pleasure." He also witnessed the power of the "varvascu," remarking that it had "leaves or roots which when thrown into the water have the faculty of intoxicating fish." As he journeyed, local inhabitants of the rainforest told him that the Amazon and Orinoco river networks were connected by waterways. He returned home without investigating this, but the theory was to be proven by the next foreign explorers allowed by the Spanish authorities into the continent in 1799, German Alexander von Humboldt and Frenchman Aimé Bonpland. They verified, with latitude and longitude, the existence of the Casiquiare Canal that connects the two vast

A lily fit for the Queen

At the beginning of the nineteenth century, two botanists exploring South America encountered the giant Amazon water lily for the first time. German botanist Eduard Friedrich Poeppig described it in 1832 as Euryale amazonica. *Then, five years later, Robert Schomburgk came across the plant in the Burbice river while he was surveying the border of British Guiana (now Guyana) for the British government. Schomburgk suggested the plant should be named after Queen Victoria. For a long time it was known as* Victoria regina *or* Victoria regia, *but plant naming rules dictate that the first published name takes priority. However, it was found that the genus of the Amazonian giant water lily differs from* Euryale, *the African giant water lily, so it is now called* Victoria amazonica. *Seeds were despatched to Kew Gardens and a plant was later sent to Joseph Paxton; it is alleged that the structure of the lily leaf later inspired his design for the Crystal Palace, which housed the Great Exhibition of 1851.*

river systems. Inspired by de la Condamine, they also made the most thorough investigation of the continent's botanical wealth thus far, collecting 12,000 specimens of plants over five years.

When Brazil gained its independence from Portugal in 1822, the previously tight controls on exploration by foreigners were relaxed. This paved the way for the English botanist Richard Spruce to explore the continent's botanical treasures. Between 1849 and 1864, he explored the Rio Negro, Orinoco, Amazon, and Andes, gathering plants to sell to sponsors in the United Kingdom. Life in the jungle was not easy, as he recalled in the letters he sent home.

LEFT *Alexander von Humboldt and Aimé Bonpland depicted by Eduard Ender during their 1799-1880 trip to Venezuela along the Orinoco river.*

RIGHT *Von Humboldt and Bonpland journeyed up the Orinoco during a five-year visit to South America in the early nineteenth century.*

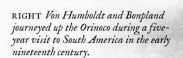

Writing in 1852 to botanist George Bentham, he complained:

"The house I am in is very old; the thatch is stocked with rats, vampires, scorpions, cockroaches and other pests to society; the floor (being simply mother earth) is undermined by sauba ants, with whom I have had some terrible contests. In one night they carried off as much farinha as I could eat in a month; then they found my dried plants and began to cut them up and carry them off. I have burned them, smoked them, drowned them, trod on them, and in short retaliated in every possible way, so that at this moment I believe not a sauba dares show its face inside the house; but they demand my constant vigilance."

In 1857, Spruce received a commission from the British Foreign and Colonial Office to collect quinine-containing cinchona specimens. At the time, quinine was the only effective treatment for malaria, and Britons beset by disease and civil disorder in India wanted to secure their own supplies rather than rely on diminishing natural sources in Ecuador. Spruce secured 637 seedlings and 100,000 seeds of the

Cassia macrophylla

from Mimoses et autres plantes légumineuses du Nouveau Continent ... recueilliés par Mess. de Humboldt et Bonpland *by Karl S Kunth, 1819.*

red bark tree, *Cinchona pubescens.* These duly gave rise to plantations in India, Ceylon (now Sri Lanka), and British Sikkim (now part of India), although the Dutch went on to dominate the trade through their plantations in Java. Spruce was ahead of his time in predicting the demise of natural supplies of botanical commodities through overexploitation. "I have seen enough of collecting the products of the forest to convince me, that whatever vegetable substance is needful to man, he must ultimately cultivate the plant producing it. Whilst the demand for such precious substances as Peruvian bark, sarsparilla, caoutchouc, &c. must necessarily go on increasing, the supply yielded by the forest will decrease, and ultimately fail."

FAR LEFT *One of the specimens collected by von Humboldt and Bonpland in South America. Scientists continue to press and dry plants for herbaria to this day.*

LEFT *Pure quinine held in the archives of Kew Gardens. Quinine-containing cinchona plants were sought after in the nineteenth century as a treatment for malaria.*

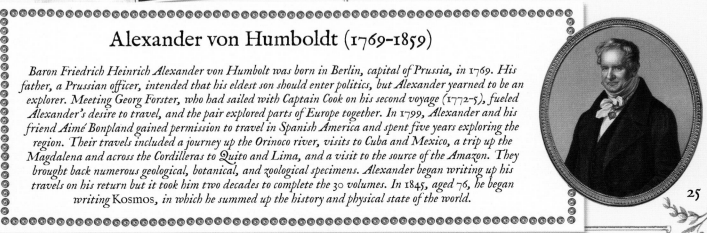

Alexander von Humboldt (1769–1859)

Baron Friedrich Heinrich Alexander von Humbolt was born in Berlin, capital of Prussia, in 1769. His father, a Prussian officer, intended that his eldest son should enter politics, but Alexander yearned to be an explorer. Meeting Georg Forster, who had sailed with Captain Cook on his second voyage (1772–5), fueled Alexander's desire to travel, and the pair explored parts of Europe together. In 1799, Alexander and his friend Aimé Bonpland gained permission to travel in Spanish America and spent five years exploring the region. Their travels included a journey up the Orinoco river, visits to Cuba and Mexico, a trip up the Magdalena and across the Cordilleras to Quito and Lima, and a visit to the source of the Amazon. They brought back numerous geological, botanical, and zoological specimens. Alexander began writing up his travels on his return but it took him two decades to complete the 30 volumes. In 1845, aged 76, he began writing Kosmos, *in which he summed up the history and physical state of the world.*

25

MASSON EXPLORES SOUTH AFRICA

Although Joseph Banks withdrew from Captain Cook's second voyage around the globe, he benefited from the journey in another way. With accommodation now available on the *Resolution*, in 1772 he sent under-gardener Francis Masson to act as Kew's first official plant collector. Masson sailed to Cape Town, then bade farewell to Cook as the ship set off for Antarctica, then known as Terra Australis Incognita, the unknown land believed at that time to lie in the far south. For the next three years, Masson roamed the Cape of Good Hope collecting seeds and plants for Banks that would make Kew the envy of other botanic gardens.

Masson's time in South Africa was spent on three separate expeditions. The first, a round trip of some 400 miles (650 kilometers), took him across the Cape Flats, to Paarl, Stellenbosch, the Hottentots Holland Mountains, and the hot springs at Swartberg and Swellendam. Traveling in a wagon pulled by eight oxen, with a Scandinavian mercenary as his guide, he got his first taste of the Cape's geographical conditions and botanical treasures, noting, "The soil of this plain is unfit for cultivation; being a pure white sand, blown by the S. E. wind from the shore of Falso Bay, and often forming large hillocks; it is nevertheless, overgrown with an infinite variety of plants peculiar to this country."

ABOVE *Francis Masson traveled hundreds of miles during his plant-collecting forays. His final trip was to North America in 1797 and he died in Montreal in Canada in 1805.*

BELOW *The Cape of Good Hope, where Masson discovered such plants as* Kniphofia rooperi, *the red hot poker.*

Brabejum stellatifolium
from Maarten Houttuyn's
Handleiding tot de plant-en
kruidkunde . . . von C. Linnaeus, 1774-83.

A flower fit for a king

Some of the most stunning plants of the South African Cape are proteas. Linnaeus named the genus after the shape-shifting Greek sea god Proteus, as it appears in many forms, from trees to low-growing shrubs. The first known example to reach Europe from South Africa was described in 1605 by Carolus Clusius in his Exoticorum libri decem *as "an elegant thistle." The most spectacular of the genus is the king protea, South Africa's national flower. Francis Masson introduced the king protea (*Protea cynaroides*) to Britain in 1775 but it was almost three decades before one flowered away from its native land. Another specimen bloomed in 1826, but the species did not flower again in the United Kingdom for 160 years, when the plant in Kew's Temperate House yielded five flowers. The plant now flowers prolifically each year.*

After returning to Cape Town in January 1773, Masson made arrangements to send back his plant collection, including many ericas that were soon thriving at Kew, before setting off in the summer on a longer jaunt. This time he journeyed with Swedish botanist Carl Per Thunberg, a one-time student of Linnaeus who was gathering plants for the Dutch East India Company. The pair traveled on horseback, accompanied by a wagonload of supplies and four helpers. Masson wrote that he was "delighted to see the luxuriance of the meadows, the grass reaching to our horses bellies, enriched with a great variety of ixiae, gladioli, and irises, most of which were in flower at the Cape in the month of August."

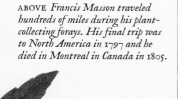

ABOVE *This potted specimen of* Encephalartos altensteinii *has thrived at Kew Gardens since Masson carried it home from South Africa in 1775.*

The terrain was not always so pleasing. The men were warned that crossing the fast-flowing river at Mostart's Hoek could prove fatal but went ahead anyway, as Masson later recalled:

"Fortifying ourselves with resolution we proceeded, and in an hour arrived at the first precipice where we looked down with horror on the river, which formed several cataracts inconceivably wild and/romantic . . . The ford was exceedingly rough, the bed of the river being filled with huge stones, which tumbel down from the sides of the mountains; but we thought our labour and difficulties largely repaid by the number of rare plants we found here. The bank of the river is covered with a great variety of evergreen trees; viz. Brabejum stellatifolium, Kiggelaria africana, Myrtus angustifolia, *and the precipices are ornamented with ericae and many other mountain plants never described before."*

A small brown folder in the archives at Kew Gardens records Masson's achievements in the Cape; in neat handwriting on lined sheets are listed 865 species, among them some 41 *Stapelia*, 8 *Amaryllis*, 86 *Erica*, 49 *Oxalis*, 47 *Pelargonium*, and 8 *Massonia*, in which the explorer's name lives on. He also went on collecting missions to Madeira, the Azores, Tenerife, the West Indies, and North America, in all introducing more than 1,000 species to the United Kingdom. His most extraordinary legacy, located in the Palm House at Kew Gardens, is a specimen of *Encephalartos altensteinii*. Possibly the oldest pot plant in the world, it has lived at Kew since Masson brought it back from the Cape in 1775. It has produced a cone only once at Kew, on which occasion Sir Joseph Banks reputedly made his last visit to the gardens.

Carl Per Thunberg
(1743–1828)

Born in Sweden, Thunberg was a student of Carl Linnaeus. He left his native country in 1770 to go to Paris and study medicine, but while away was asked if he would be willing to travel to Japan to gather plants and seeds on behalf of a wealthy Dutch collector. At that time, Japan allowed entry only to employees of the Dutch East India Company, so it was arranged for Thunberg initially to enter the service of the company at the Cape in South Africa so that he could perfect his Dutch. Arriving in 1772, he was to stay for three years; it was during this visit that he explored the interior with Francis Masson. He eventually traveled to Japan as a doctor. During his stay, he amassed large numbers of specimens, which he described in his most important work, Flora Japonica.

A PLETHORA OF PLANTS FROM DOWN UNDER

The various indigenous peoples of Australia were the first true plant collectors on the continent, gathering a wide number of species for food, medicines, and tools. They came to have a highly developed understanding of the native flora, and developed their own naming systems, which were often more discriminating than European ones. However, much of the aboriginal verbal tradition has been lost since outsiders arrived, or is only slowly being recovered. The first verified collector to introduce plants from Australia to other areas of the world was William Dampier. In 1699, he gathered specimens of 24 species from Shark Bay and the Dampier Archipelago of Western Australia. He described a tree that is now considered to be the native willow (*Pittosporum phillyraeoides*) and he also mentioned a grain that grew on bushes that is now believed to be the bird-flower (*Crotalaria cunninghamii*).

The first extensive collection of Australian flora to arrive in Europe was that collected by Sir Joseph Banks and Daniel Solander on Cook's first three-year circumnavigation between 1768 and 1771. The expedition traveled around New Zealand before sailing up Australia's east coast from south to north. On reaching one particularly rich floral site, Banks persuaded Cook to name it Botany Bay. "Our collection of Plants was now grown so immensly [sic] large," he wrote, "that it was necessary that some extrordinary [sic] care should be taken of them least they should spoil in the books [in which they were dried]." In 1788, the year in which Sydney was founded, nurserymen Lee and Kennedy proudly displayed five new Australian species. These included *Banksia serrata*, which was named after Sir Joseph, and claimed to be the first plant grown in England from Botany Bay seeds.

The York Road poisoner

The Scot James Drummond was curator of the botanical garden of the Royal Cork Institution in Ireland, before emigrating to the fledgling Swan River colony of western Australia in 1829. During his time there, he collected plants for botanists at Kew Gardens and other institutions. In the early 1830s and 1840s, many sheep, goats, and cattle died from an unknown cause, and Drummond determined to help find the reason. A goat fed the juice of one particular leguminous plant appeared languid the following morning and was fed a second dose. It cried out and died soon after. The plant that had poisoned the goat, Gastrolobium calycinum, was one of a group of poisonous peas that were later found to have been affecting the livestock. Gastrolobium calycinum is still colloquially known as York Road Poison, after the location at which many of the Swan River colonists' animals met their fate.

ABOVE *Robert Brown traveled as the army surgeon and botanist on Matthew Flinders's expedition to Australia.*

BELOW *Dampier mentioned a "grain that grew on bushes" that may have been* Crotalaria cunninghamii, *which is reproduced here from* Curtis's Botanical Magazine, 1869.

By 1800, Banks was President of the Royal Society, and England was at war with France. Despite the hostilities, the Institut National in Paris approached Banks requesting safe passage to Australia for two ships under the command of Nicolas Baudin "to continue the useful discoveries which your navigators made in their voyages around the world." Banks recommended that permission be granted and the ships set sail in October, with eight naturalists, three artists, and five gardeners among those on board. A British expedition launched soon after, headed by Matthew Flinders and including the army surgeon and botanist Robert Brown, plus artist Ferdinand Bauer. It traveled with the dual purpose of exploring unknown territory and keeping an eye on the French.

Banksia serrata
from Henry Charles Andrew's
Botanist's Repository, Volume II,
1790s-1830.

vegetable," he wrote. Cunningham spent 15 years wandering across Australia on numerous expeditions, having decided that he could "blend discovery with botanical research tolerably well." He visited the northeast's tropical forests, the northern mangroves, arid stretches of the western coast, the Illawarra's rainforests, plus Timor, Van Diemen's Land, and New Zealand. When he returned to England in 1831, his ship was forced to shelter from a gale in Sydney Harbour. Cunningham went ashore and found an orchid specimen he had been looking for for ten years. In all, he collected some 1,300 species. Cunningham returned to Australia in 1837 to take on the role of Colonial Botanist. After a trip to New Zealand he became ill with tuberculosis, concluding just before he died: "I can neither undertake any more expeditions nor walk about in search of any more plants."

RIGHT Pittosporum phillyraeoides *from J E Brown's* The Forest Flora of South Australia, 1882-93.

Arriving in Cape Leeuwin, the French expedition traveled north up Australia's west coast before retracing its steps and starting to survey the south coast. Meanwhile, Flinders traveled east from Cape Leeuwin along the south coast. Eventually, the French ship caught up with the English, raised a flag of truce, and the two teams had an "agreeable meeting." Baudin died before making it back to Europe, but the Muséum national d'Histoire in Paris hailed the expedition's collection as "the richest ever received." Flinders, having tried to return to England on a different ship, was detained in Mauritius by the French governor for six and a half years. His team returned home in 1805, carrying 4,000 species of plants, 1,700 of which were new to science. Robert Brown sent seeds of *Bossiaea dentata* back to Kew and also introduced *Solanum hystrix* (Afghan thistle); meanwhile Peter Good, the gardener on the same expedition, was credited with introducing *Eremophila glabra* (common emu bush) to Kew Gardens in 1803. However, their success was eclipsed by the news of Nelson's victory and death a few hours earlier.

In 1814, Kew Gardens' head gardener William Townsend Aiton wrote to Banks, urging him to send botanists abroad once more. William Kerr had been the last botanist sent abroad from Kew before hostilities with the French thwarted Banks's collecting plans. The arrival of a definitive treaty with France, coupled with concern that Austria might create a garden superior to Kew at Schönbrunn, prompted Banks to despatch Allan Cunningham to New South Wales and James Bowie to the Cape of Good Hope, via South America. "The plants of both these countries are beautiful in the extreme and are easily managed as they suit the conservatory and have no occasion for the unnatural heat required by the intertropical

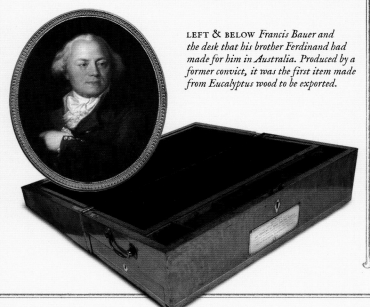

LEFT & BELOW *Francis Bauer and the desk that his brother Ferdinand had made for him in Australia. Produced by a former convict, it was the first item made from Eucalyptus wood to be exported.*

A modern plant-collecting heyday

Although early visitors to western Australia were voracious collectors, contemporary searches are yielding new species at an unprecedented rate. Scientists estimate the region's current flora to be 13,000 species, with 6,000 in the temperate south. Of this 6,000, 75 to 80 percent are endemic (i.e. not found anywhere else). Professor Stephen Hopper, the present Director of Kew Gardens, has estimated that as many as 25 percent are still undescribed, so the numbers are likely to increase in the coming years. The rate of discovery is exemplified by the Anthericaceae family. When it was last documented in 1986, it was considered to have 97 species. Since then, new discoveries have expanded the family by almost 20 percent, to 115 species.

THE WILDS OF NORTH AMERICA

'At the start of the nineteenth century, the United States of America was a fledgling country. The 13 states that had originally been British colonies had fought a war for independence against Britain from 1775, and in 1783 Britain had been forced to recognize the colonists' sovereignty. The original 13 east-coast states were subsequently joined by Vermont in 1791, Kentucky in 1792, Tennessee in 1796, and Ohio in 1803. However, the French still owned the largely unexplored area of Louisiana to the west of the Mississippi river, the Spanish laid claim to land beyond it, and the British were keen to expand into northwest America. In 1793, the Scot Alexander Mackenzie, working for the North West Company, crossed from the Atlantic to the Pacific from present-day Alberta to the Strait of Georgia, reporting the route to the English Crown. This prompted the newly inaugurated third President of the United States, Thomas Jefferson, to launch an expedition to find a route from the Mississippi river to the Pacific.

By the time the Corps of Discovery expedition left civilization for the Pacific in 1804, the U.S. had purchased Louisiana. Expedition leaders Meriwether Lewis and William Clark and their team were now exploring unknown U.S. territory rather than trespassing on French land. Jefferson instructed them to record the names and geographical details of the indigenous North American nations, to map the route, noting the soil types, animals, minerals, fossils, and climatic zones they encountered, and record "the dates at which particular plants put forth or lose their flower, or leaf, times of appearance

ABOVE *Meriwether Lewis was eventually made governor of the Louisiana Territory. There is some mystery as to whether his death in 1809 was suicide or murder.*

RIGHT *William Clark, a former soldier with the U.S. Army, became governor of the Missouri Territory after his expedition with Meriwether Lewis.*

of particular birds, reptiles or insects." Lewis was to be the expedition's naturalist and Clark its cartographer.

By July 1804, the expedition had traveled from St. Louis on the Mississippi river on to the Missouri river and deep into the prairies. Here, they collected specimens of prairie wild rose (*Rosa arkansana*), silver-leaf Indian breadfruit (*Pediomelum argophyllum*), and skunkbush sumac (*Rhus trilobata*). In his journal, Clark enthusiastically noted, "the praries Contain Cheres, Apple, Grapes, Currants, Rasp burry, Gooseberris, Hastlenuts and a great Variety of plants & flours not Common to the US. What a field for a Botents [Botanist] and a natirless [Naturalist]." When they reached the high plains of today's North Dakota, they overwintered at the Mandan villages among a group of 7,500 Indians. The Indians traded corn, squashes, beans, and roots for tools and other items.

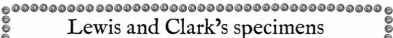

Lewis and Clark's specimens

Some of the plants collected by Lewis and Clark were lost on the journey. Those that survived were given to the highly regarded German botanist Frederick Pursh so that he could draw and arrange them. Later in his life, Pursh wrote Flora Americae Septentrionalis; or, a Systematic Arrangement and Description of the Plants of North America. *He included 132 plants found by Lewis and Clark and proposed 94 new names that were at least partly based on material they gathered. Forty of the new names he proposed are still in use. Had Pursh not written this tome, the Lewis and Clark plants would probably not have been used to describe new species and the pair might not have received any credit for locating them.*

David Douglas (1799–1834)

Born in the village of Scone, Scotland, David Douglas worked at the Botanical Gardens in Glasgow before joining the Horticultural Society of London. Unable to visit China because of the unstable political situation of the 1820s, he was dispatched to New York. He gathered a wide range of ornamentals plus new varieties of fruit before returning to Britain a year later. On a second trip to North America, he stayed three years. It was during this trip that he encountered the fir that now bears his name, noting that it "exceeds all trees in magnitude." On a further trip, he headed to California. Here, he encountered 360 species and 20 genera previously unknown to the West, including the noble fir (Abies procera). He came to a grisly end in Hawaii when he fell into a cattle pit and was gored to death by a bullock.

BELOW *Also known as the Monterey pine,* Pinus radiata *is native to coastal California.*

The next section of the journey took them to the Missouri river's Great Falls, where they noted plants such as the narrow-leafed cottonwood and the tree we now know as the Douglas fir. After resting for a couple of days at what is now Lolo, Montana, they began to cross the Rocky Mountains, passing through, as Clark recorded it, "a thickly timbered Countrey of 8 different kinds of pine, which are So covered with Snow, that in passing thro them we are continually covered with Snow, I have been wet and as cold in every part as I ever was in my life . . ." Nonetheless, they made it across and in autumn 1805 they reached the tidal waters of the Columbia Estuary. With the Pacific finally reached, Clark wrote, "Ocian in view, O! the joy." The expedition proved that there was no interior Northwest Passage through the mountains, and gathered many plant specimens new to science along the way.

While Lewis and Clark were making their journey west across America's *terra incognita*, Sir Joseph Banks and a collection of leading academics had founded the Horticultural Society of London, now the Royal Horticultural Society. In 1823, the Society sent David

Douglas to New England on the northeast coast so that they, too, might learn something of this vast continent's botany. Despite having his horse bolt while out riding, being robbed of his belongings, and almost sinking with his boat in a storm, he came back with a wide range of ornamentals plus new varieties of apple, pear, plum, peach, and grape. The Society hailed his trip as a "success beyond our expectations" and sent him off again six months later to explore the wilds of the Pacific northwest.

It was during this trip that he first encountered the fir that Lewis and Clark had noted 20 years earlier, and which now carries Douglas's name. He wrote that it "exceeds all trees in magnitude. I measured one lying on the shore of the river 39 feet in circumference and 159 feet long; the top was wanting . . . so I judge that it would be in all about 190 feet high." During his three-year stay, Douglas covered 7,032 miles (11,317 kilometers) of ground before returning to the United Kingdom. Two years later he was back, but this time his luck ran out. Having gone to Hawaii to recuperate after losing his plant collection and journal in a canoeing accident, he met a tragic end at the age of only 35. During his short life, he introduced more than 200 species to the U.K., including many "Californian annuals" and a number of evergreens that are now widespread.

LEFT Rosa arkansana *from Britton & Brown's* An Illustrated Flora of the Northern United States . . . , 1896.

RIGHT *It was here, at the Missouri Falls, that Lewis and Clark spotted the tree that later became known as the Douglas fir after David Douglas introduced it to the U.K.*

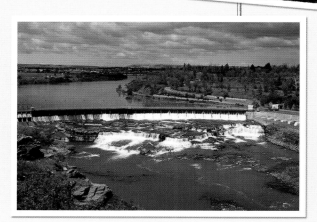

JOSEPH HOOKER IN THE HIMALAYA

Born in Suffolk, England, in 1817, Joseph Dalton Hooker was influenced by his father's love of botany from a young age. During Joseph's early years, William Jackson Hooker held the chair of botany at Glasgow University; he later became director of Kew Gardens. He observed that at the age of 15, his son was "contented and happy at home and studying Orchideae most zealously."

Joseph's first taste of foreign plant collecting came when he joined James Clark Ross's voyage to Antarctica in 1839 as assistant surgeon and botanist on the ship *Erebus*. He was instructed to collect representative species at each location and look out for potentially valuable plants, such as Antarctic lichens that might yield dyes, and tree ferns on St. Helena that could provide a substitute for hemp.

ABOVE *Joseph Hooker, seen here sitting in the center with the light-colored hat on, traveled widely. This photograph was taken while on a camping trip during a visit to the United States in the 1870s.*

The four-year expedition, which aimed to determine the position of the magnetic south pole, called at Madeira, the Cape of Good Hope, Tasmania, New Zealand, Australia, the Falkland Islands, and the tip of South America, giving Hooker ample opportunity for plant hunting. He reveled in arriving at new places and making new botanical discoveries.

He considered *Pringlea antiscorbutica*, a cabbage-like plant that had been found to prevent scurvy, " . . . the most interesting plant procured during the whole voyage . . . The contemplation of a vegetable very unlike any other in botanical affinity and in general appearance, so eminently fitted for the food of man, and yet inhabiting one of the most desolate and inhospitable spots on the surface of the globe, must equally fill the mind of the scientific explorer and common observer with wonder."

It is for his plant gathering in Sikkim, in India, that Hooker is best known. He left for India in 1847, arriving in Darjeeling in 1848. His plan was to head for the high mountains of Sikkim but he was forced to linger awhile in Darjeeling. With relations between Britain

BELOW *A view of the Himalaya that appeared in the second volume of Joseph Hooker's* Himalayan Journals *in 1854.*

and Sikkim frosty, the local Rajah was not keen to grant him entry to the region. Hooker passed the time studying the local ferns, lichens, mosses, orchids, arums, magnolias, and rhododendrons.

On a trek to the Great Rungeet (Rangit) river 11 miles (18 kilometers) to the north, he encountered three rhododendrons, one of which, *Rhododendron dalhousiae*, he described as: "a parasite on gigantic trees, three yards high, with whorls of branches, and 3–6 immense white, deliciously sweet-scented flowers at the apex of each branch." He considered it "the most lovely thing you can imagine."

He eventually set out for Sikkim in October 1848, almost a year after leaving Britain. Not one to travel light, he was initially accompanied by 56 locals who acted as porters, collectors, and guards. The party ascended high peaks and dropped

Naming names

Hooker named several of the plants he collected after people who had helped him during his Indian expeditions. Rhododendron dalhousiae *was named after Lord Dalhousie's wife,* Magnolia campbellii *honored Archibald Campbell, and* Hodgsonia heteroclita *immortalized Bryan Hodgson, a scholar and zoologist with whom Hooker stayed in Darjeeling and "in whose hospitable residence my examination of this splendid plant was conducted."* Cathcartia villosa, *now known as* Meconopsis villosa, *carries the name of James Ferguson Cathcart, a civil servant sent to work in India. He met Hooker toward the end of his career, when he was living in Darjeeling and employing local artists to sketch the Himalayan flora. He gave the 1,000 or so pictures to Hooker, and the artist Walter Hood Fitch used them to prepare paintings for Hooker's 1855 book* Illustrations of Himalayan Plants.

into deep valleys, often having to cross fast-flowing rivers on precarious bridges. However, Hooker considered the collecting opportunities worth the dangers; once back in Darjeeling, he was able to send back 80 porter-loads of specimens to Kew.

After overwintering in Darjeeling, Hooker left on a second expedition in May 1849. However, his progress was hindered by the Dewan (prime minister) of the Rajah, who did not want him back in the country. Nonetheless, Hooker collected ten types of rhododendron including two new species, *R. griffithianum* var. *aucklandii* and *R. edgeworthii*. He considered that these "in the delicacy and beauty of their flowers . . . perhaps excel any others."

In October, Hooker met up with Archibald Campbell, the political agent to Sikkim who mediated between the Rajah and the British government. The pair traveled into Tibet and then on to eastern Sikkim to visit the Chola and Yakla passes. The trip was successful, Hooker gathering 24 rhododendron species and also encountering *Cathcartia villosa*, the Himalayan woodland-poppy, but they were turned back from the Chola Pass.

That night, Campbell was attacked by a group of men and taken prisoner at the orders of the Dewan. Hooker remained with his friend and eventually got a letter through to Lord Dalhousie, Governor-General of India, with whom Hooker had stayed in Calcutta. Dalhousie sent an English regiment to the border and planned an invasion, although that never happened. As a result of the incident, southern Sikkim was annexed to India, in effect adding to the British Empire. This land was very fertile and it was to here that Britain later transferred cinchona from South America and tea from China.

Rhododendron dalhousiae
illustrated by Walter F. Fitch (1817–92), from the archives of the Royal Botanic Gardens, Kew.

Cathcartia villosa
from Curtis's Botanical Magazine, 1851.

ABOVE *A tea brick that Hooker brought back with him from India.*

RIGHT *Joseph Hooker was the Director of Kew Gardens for 20 years and he is buried in the churchyard of St. Anne's on Kew Green.*

ABOVE RIGHT *Joseph Hooker's passport from Sikkim. It was the first time that the Lepcher script had been seen in written form by Western eyes.*

RIGHT *The teapot that Hooker took with him on his travels to the Himalaya.*

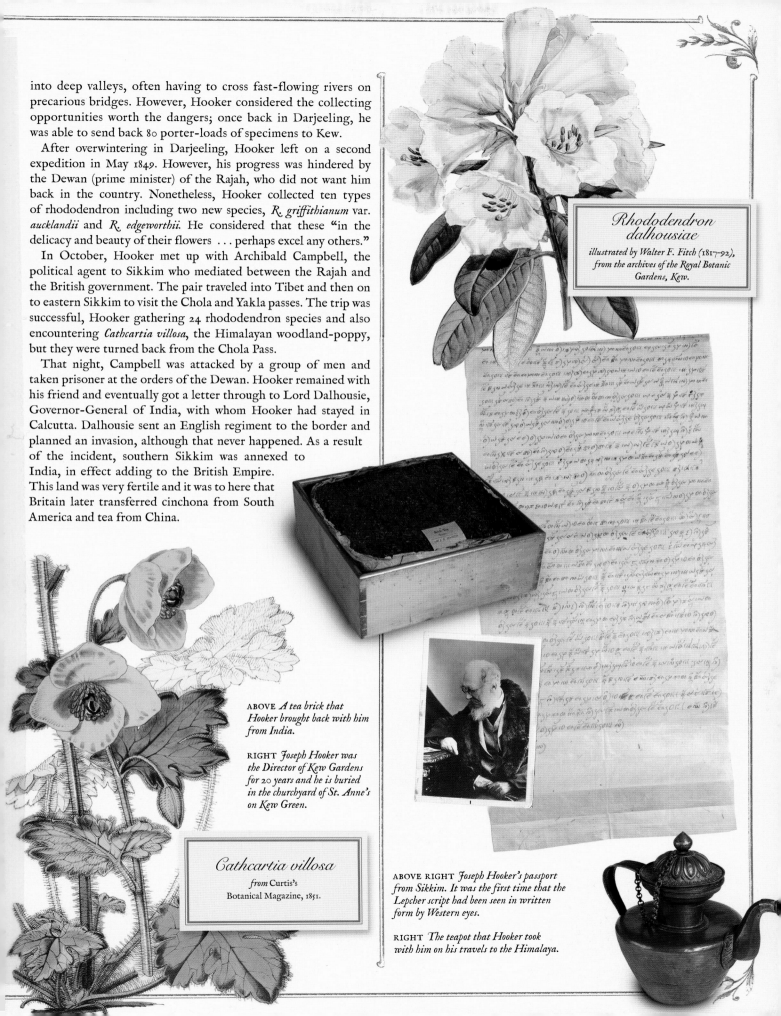

EXPLORING THE TREASURES OF THE ORIENT

A single dried specimen of the handkerchief tree (*Davidia involucrata*) was to steer 23-year-old Ernest Henry Wilson away from a career teaching botany to a life of plant hunting. The sample was sent by Dr. Augustine Henry, an Irish doctor working in China, to Kew Gardens in 1891. After seeds that Henry sent failed to germinate, he urged Kew to send a collector to gather samples. James Veitch, the proprietor of Veitch nurseries, felt that the plant would be popular with the emerging breed of suburban gardeners and decided to sponsor a collector. At the suggestion of Kew's director, William Thistleton-Dyer, Veitch selected Wilson. He gave him a three-year contract to seek out and bring back *Davidia involucrata*, instructing him not to "dissipate time, energy, or money on anything else."

Wilson arrived in Yichang, the port on the Yangzi river that would be his base for two years, in February 1900. Dr. Augustine Henry had agreed to help him locate the *Davidia* but his directions amounted to little more than a cross on a hand-drawn map covering 20,000 square miles (50,000 square kilometers). Undaunted, Wilson gathered together a team of men to carry his equipment and set off. When they eventually arrived in the area where the tree was located, a local guide agreed to take them to the exact spot. Instead of seeing a beautiful tree, however, they encountered a stump and the newly built wooden house that had been fashioned

ABOVE E. H. Wilson. Over 60 species of plant bear his name.

from it. Fortunately, Wilson encountered another of the trees in full bloom while collecting southwest of Yichang. He noted that its white flowers were akin to "huge butterflies or small doves hovering amongst the trees."

When Wilson returned to England, Veitch was pleased with his work. Not only had he brought back seeds of the *Davidia*, but he had also collected many other seeds and bulbs, as well as compiling a herbarium of 2,600 records. By January 1903, newly married Wilson was China-bound on Veitch's behalf once more, this time to gather *Meconopsis integrifolia* from the uplands of Tibet. Despite almost having his boat ripped apart on jagged rocks amid the Yeh-Tan rapids, Wilson made it to Tibet and found the desired golden fields of *Meconopsis*. During two forays into the mountains, he collected seeds of this and *M. punicea*, plus many more specimens besides, writing to Thistleton-Dyer, "On these two trips I have collected specimens of some 900 species of plants, a complete set of which I hope will eventually find a home in Kew Herbarium."

Once he was back home, Harvard University's Arnold Arboretum persuaded Wilson to go on a third collecting trip to China on its behalf. This time he was to "increase the knowledge of the woody plants of the [Chinese] Empire and to introduce

Meconopsis integrifolia
from Curtis's Botanical Magazine, 1905.
It can grow at altitudes of 11,000–15,500 feet (3,500–4,720 meters).

Meconopsis punicea
from Curtis's Botanical Magazine, 1907.
It is also known as red poppywort and can be found in western central China and the mountains of Tibet.

into cultivation as many of them as is practicable". Despite contracting malaria, Wilson managed to gather *Acer wilsonii*, *Clematis tangutica* subsp. *obtusiuscula*, the dogwood *Cornus kousa* var. *chinensis*, and *Rhododendron moupinense*. Wilson had a soft spot for rhododendrons, later writing, "To traverse the mountains of western China in the rhododendron season is to enjoy a feast of beauty not excelled the world over."

Once the trip was over, Wilson brought his family to Boston and took up a temporary position at the Arnold Arboretum, supervising the organization of his herbarium collection. However, it was not long before he was off again, this time to gather bulbs of the regal lily (*Lilium regale*) that he had encountered on an earlier trip. Eventually, he made it to the remote Min valley, where "in summer the heat is terrific, in winter the cold is intense, and at all seasons these valleys are subject to sudden and violent windstorms against which neither

RIGHT *Rhododendrons in bloom at the Arnold Arboretum, part of Harvard University.*

Rhododendron moupinense

from Curtis's Botanical Magazine, 1915.

man nor beast can make headway." After marking the position of 6,000 bulbs that were to be lifted in October, Wilson was being carried in his sedan chair on a mountain path when a landslide struck. He was injured in one leg, leaving him with what he called his "lily-limp."

Wilson made two more trips, during which he explored Japan, Korea, and Taiwan with his wife and daughter. In Japan he visited the city of Kurume in Kyushu, where he saw a century-old collection of 250 named azaleas. He brought a selection of them back to North America, saying, "Proud am I of being the fortunate one to introduce this exquisite damsel to the gardens of eastern North America." The collection became known in England as "Wilson's 50," though in fact there were 51. Wilson returned to work at the Arnold Arboretum after his travels but died prematurely in a car crash with his wife in 1930. In his lifetime, he had introduced over 1,000 species to Western gardens, the majority from China. In his book *Plant Hunting,* he wrote, "To no part of the world do gardens owe more than to China—the Kingdom of Flowers."

Azaleas from the sacred mountain

Wilson's 51 Kurume azaleas were displayed in March 1920 at the Orchid Show hosted by the Arnold Arboretum. Aside from a showing of 12 varieties at the Panama-Pacific Exposition in San Francisco in 1915, this was the first time they had been seen in America. The owner of the garden from which Wilson had sourced the azaleas, Kijiro Akashi, explained that they originated from a Japanese gentleman named Motozo Sakamoto, who had lived in the city of Kurume a century before Wilson's visit. "The gardens of Messrs Akashi and Kuwano," said Wilson, "the two leading experts, were veritable fairy-lands, and I gasped with astonishment when I realized that garden-lovers of America and Europe knew virtually nothing of this wealth of beauty."

Clematis tangutica

from Revue Horticole, 1834-56. *It is commonly called golden clematis.*

THE COST OF SUGAR CANE

*H*umans have always had a sweet tooth. In ancient times, the Chinese used the sap from sugar palm trees (*Arenga pinata*—synonym *A. saccharifera*) as a sweetener, while Africans relied on sweet sorghum. In New Guinea, the locals were partial to the sweet juice from sugar cane. When they traveled across the sea 8,000 years ago to Indonesia and the Philippines, they took canes with them and began a chain of events that would see sugar rise to dominate all other sweeteners. Today, we devour 160 million tonnes each year, in everything from breakfast cereals to carbonated drinks.

From the Pacific Islands, sugar cane eventually reached the northern tip of India. Charaka, an Indian physician who lived in the first century B.C., wrote in the medical text *Charaka-Samhita*, "The juice of the sugar-cane, if the stalk is chewed with the aid of the teeth, increases the semen, is cool, purges the intestines, is oily, promotes nutrition and corpulence, and excites the phlegm." Historians believe that it was in India that communities advanced from extracting and boiling cane juice to making crystallized sugar from it.

The knowledge spread from India to Persia around A.D. 600 and was then carried further west by the Muslim Arabs. When the Crusaders set out to reclaim the Holy Land from Islam from the eleventh century, they learned sugar-making techniques and started producing the commodity themselves in Cyprus, Crete, Rhodes, and Greece. From the fifteenth century, however, the discovery of new lands began to shift the heart of the trade west. Portugal's Prince Henry the Navigator set up a sugar mill on the island of Madeira shortly after its discovery by João Gonçalves Zarco in 1420. In 1444, he shipped 235 Negroes from Lagos, Nigeria, to the island to work in the cane fields, the first slaves to be bought in from Africa. Soon, Madeira was the largest sugar producer in the West, generating 1,790 tons per year.

ABOVE *A detail of the stalk of* Saccharum officinarum.

Arenga saccharifera
known as sugar palm, from
Robert Bentley and Henry Trimen's
Medicinal Plants, 1875-80.
The Chinese used it as a sweetener.

One of the island's customers was Christopher Columbus. Sent to buy 2,400 arobas (79,000 pounds/36,000 kilograms) of sugar, he ended up marrying an island girl and set up home there. After he made his fabled crossing of the Atlantic in 1492, he introduced sugar cane to Hispaniola, whence the industry spread to São Tomé, Brazil, Cuba, Jamaica, and Mexico. Having gained exclusive rights to the South American continent under the Treaty of Tordesillas, the Spanish and Portuguese were to dominate the sugar trade from then until the latter part of the seventeenth century.

In the latter decades of the sixteenth century, the Dutch, British, and French switched from being transporters and consumers of sugar to wanting their own piece of the action. The Dutch took over coastal areas of Brazil, while the British and French snatched Caribbean islands. British-owned Barbados became one of the major suppliers. A triangular trade was established whereby British goods were bartered for slaves in Africa. The slaves were shipped to the West Indies in inhuman conditions, and the sugar they toiled

BELOW *A painting of a sugar mill in Antigua that was made by William Clark in 1823. The wind would power the crusher that squeezed the juice out of the cane.*

The plants that give us sugar

Sugar cane has a complicated and poorly understood genealogy. Two cultivated species have a limited distribution: Saccharum barberi *in northern India and* Saccharum sinense *in China. The sugar cane of modern commerce (the "noble canes") is* Saccharum officinarum. *Its wild ancestor is probably* Saccharum robustum, *first taken into cultivation in the region of Papua New Guinea. Sugar cane spread out across the Pacific Islands and into Asia, then the Mediterranean. Columbus took sugar canes—possibly a hybrid of* Saccharum barberi *and* Saccharum officinarum—*to Hispaniola in 1493 and this variety was to become the basis for the New World sugar industry.*

to produce was sent to Britain. Over the course of 350 years, between 10 and 15 million Africans were shipped to the New World to help satisfy rising demand for sugar.

On the plantations, sugar cane was simultaneously planted, cut, milled, and boiled. The cane was crushed in a mechanical mill, after which the extracted juice was taken to the boiling house. Here it underwent boiling in several cauldrons of diminishing size. Toward the end of the process, the viscous syrup had to be stirred to prevent it from burning. The Portuguese Jesuit Father Antônio Vieira, in a sermon in Bahia in 1633, likened the sight of men toiling at the cauldrons amidst eternal flames and smoke to a vision of souls in Hell.

In the late eighteenth and early nineteenth centuries, the slave trade was finally abolished and the New World sugar cane industry went into decline as

Saccharum officinarum
from François Regnault's
La botanique mise à la portée de
tout le monde, 1774.

plantation owners struggled to instigate new forms of labor, fund new steam-powered mills, and compete with newly discovered sugar beet. In 1750, the Jesuits carried the cane to North America and in 1788, sugar cane traveled from England to Australia on board the First Fleet convict ships bound for Botany Bay. The plantations there employed local Polynesians from nearby islands.

After departing from New Guinea and traveling the globe, the world's favorite sweetener was finally cultivated by the race who had set it on its journey thousands of years before.

Traditional uses for sugar, such as making rum and cachaça, are now being supplemented by producing ethanol from molasses for fuel. More than half the world's ethanol production is made from sugar and sugar by-products, with Brazil being the world leader. Eighty percent of new cars now sold in Brazil are equipped to use ethanol as well as gas. As climate change takes hold and our need for non-fossil sources of fuel increases, the ethanol market is set to expand. Our love affair with sugar is far from over.

Sugar beet and sugar cane

In 1747, Andreas Sigismund Marggraf discovered that sugar also existed in the plant Beta vulgaris, *which could be grown at more northerly latitudes than cane. This gave rise to a rival sugar beet industry, which for a while took trade from sugar cane. However, when scientists realized that some varieties of sugar cane set seed, rather than being sterile and having to be reproduced vegetatively as traditionally had been the case, they began breeding new varieties. It is now possible to grow varieties that are resistant to disease, favor certain soil, water, and climatic conditions, or germinate more quickly to reduce weed growth. Today, after declining in the nineteenth century, sugar cane accounts for three-quarters of the global market, and sugar beet makes up the remaining quarter.*

ABOVE *A loaf of sugar from Kew's archives. It was made from canes grown in England. Sugar loaves got their name from the fact that their shape was similar to that of the Sugar Loaf Mountain in South Africa.*

BELOW *Slaves cutting sugar cane in the West Indies c. 1870.*

The Trading Company That Ruled India

The "Company of Merchants of London Trading into the East Indies" came into being on the last day of the year 1600. The company was given a Royal Charter by Queen Elizabeth I of England, giving it a monopoly on trade with the East Indies, an area loosely incorporating South and Southeast Asia. Other countries founded similar companies during the seventeenth century: the Dutch set up Vereenigde Oost-Indische Compagnie or VOC in 1602; the French later established La Compagnie française des Indes orientales; while Sweden set up Svenska Ostindiska Companiet.

Gossypium herbaceum
also known as Levant cotton,
from the archives of the Royal Botanic
Gardens, Kew. It spread from Africa and
Arabia across Asia to China.

With spices being the "must-have" luxuries of the day, the East India Company's first voyage set off for the pepper-producing islands of Sumatra and Java in 1601. The four ships' crews tried trading woolen fabrics from England for spices, but found little demand for them in such hot, tropical climes. So, expedition leader James Lancaster captured a Portuguese ship and used its cargo of gold, silver, and Indian textiles to buy 500 tons of pepper. Lancaster set up a trading post or "factory" at Bantam, Java, before returning to England. Over the next nine years, the company launched 11 more trading journeys to Bantam.

The British soon realized that Indian cotton would give them much greater bartering power than English woolens and began setting up factories on the Indian mainland. The first of these, in 1615, was at Surat, on the northwest coast, an area at the time administered by the Mughal Empire. By 1690, the Company had set up trading centers all along the west and east coasts of India. Those at Madras, Calcutta, and Bombay became the most important and, over time, they evolved into major commercial towns under British jurisdiction. The Company defended its stations with its own military forces. After defeating French ambitions for colonial territory in India, the Company began overpowering local rulers. By 1833, the Company's domain had become vast.

The Company had been regularly trading with China since 1699 and in the subsequent century tea had become an increasingly popular drink in Britain, and thus a lucrative product. In the eighteenth century, demand for tea was rising in Europe but there was little that China wanted in exchange from the West for its precious commodity. The Manchu emperor Qianlong made this clear in a letter he wrote to King George III, saying, "I set no value on strange objects and ingenious, and have no use for your country's manufactures." China therefore demanded payment for tea in silver bullion.

The East India Company had the monopoly on importing tea to Britain, but because taxation on tea was high, smuggling was rife, and acquiring silver became increasingly difficult. By 1773, war and famine in India, plus an economic downturn in Europe, conspired to push the Company's finances towards collapse. To avoid this,

ABOVE *A painting of the Dutch East India Company fleet returning to Amsterdam in 1599.*

BELOW *Fort St. George, in Madras, depicted in the eighteenth century when it was an important British trading center.*

RIGHT *Boston citizens, dressed as Native Americans, board English ships and throw tea chests into the harbor in protest at the East India Company undercutting local merchants.*

RIGHT *Opium pods that have been striated in order to extract the juice.*

the British government passed the 1773 Tea Act, allowing the Company to sell its tea directly to North America at a lower rate of duty. The colonists were angered by this, as it allowed the Company to undercut local merchants, and they eventually took action by throwing 342 crates of East India Company tea into Boston harbor. This helped spark the American Revolution, at the end of which the 13 British colonies in North America gained independence to form the United States of America.

Although the colonists now began making their own deals with China, the Company continued to have a monopoly on British trade with the Orient. However, it found it increasingly difficult to raise the silver demanded by the Chinese in exchange for tea, so it began selling Indian-grown opium to Chinese merchants in exchange for bullion. The illegal drug was then smuggled into China. By the early nineteenth century, people in Britain were becoming uneasy that the Company was juggling the dual roles of controlling trade with the East and governing much of India. One spokesperson wrote,

Papaver somniferum
from the King's College Collection held at the Royal Botanic Gardens, Kew.

"We object to their [the Company] being allowed to combine in their own persons the separate and irreconcilable functions of tea-dealers and rulers of a mighty empire."

The government took heed and in 1833 abolished the company's trading rights in favor of allowing it to continue to administer British India on behalf of the Crown. Because the Company was no longer dependent on Chinese supplies of tea, it hatched a plan to transplant seedlings from China to India. An Indian tea industry was soon thriving, and the Chinese monopoly was smashed. All was not well on the subcontinent, however. Rebellions by Indians serving in the Company's armed forces in 1857 set the country on the path to independence and prompted the British government to nationalize the Company. After nearly three centuries of trading in the plant-based commodities of spices, cotton, tea, and opium, the East India Company was officially dissolved in 1874.

BELOW *Smokers in an opium den. The East India Company smuggled opium from India to China to raise silver bullion with which to buy tea.*

COLONIAL ENDEAVORS

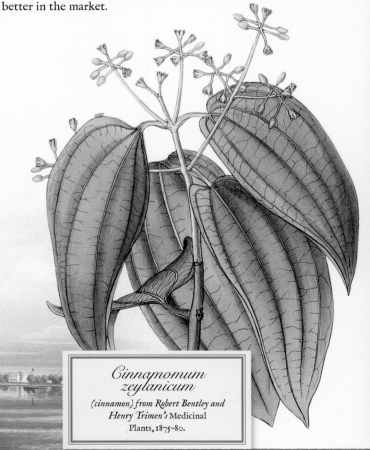

As Western nations continued to explore new parts of the world during the early eighteenth century, exotic plant species flooded into Europe. Initially, botanic gardens concentrated their efforts on growing and showcasing these curiosities at home. However, keeping plants from the tropics alive on long sea journeys and in the more temperate climes of their new residences could be difficult. Kew Gardens operated at least two "hospital hothouses," in which its gardeners struggled to keep alive sickly newcomers. In the 1770s, Pierre Poivre, a missionary naturalist turned colony administrator, had the bright idea of obtaining spice-yielding plants from the Moluccas and growing them on Île de France (now Mauritius), a French colony, rather than in Paris's botanical garden. Initially, he brought five nutmeg plants from the East; these were later supplemented by "20,000 nutmegs, as seeds or plants, and 300 clove trees." As a result, the Dutch monopoly on the spice trade was broken almost overnight.

This idea of growing economically lucrative plants in the more suitable environments of colonial gardens was soon embraced by Britain. Colonel Robert Kyd suggested to his employer, the East India Company, that a garden located close to Calcutta's port could supply teak to the Navy for shipbuilding. He also envisaged introducing cotton, tobacco, coffee, tea, indigo, sarsaparilla, sandalwood, pepper, cardamom, camphor, nutmeg, and cloves to new territories. The garden was founded in 1787. From Kew Gardens, Sir Joseph Banks advised the company about which other economic plants it should seek out and cultivate.

Banks recognized the potential economic benefits of exchanging plants between gardens in the colonies and envisaged that Kew might become "a great botanical exchange house for the empire." He was soon involved with setting up new gardens

in the West Indies and transferring plants to them. When he sent Captain Bligh on the ill-fated *Bounty* expedition of 1787, it was primarily to transfer breadfruit from Tahiti to St. Vincent. The hope was to obtain a reliable food-stock for slaves and therefore boost the output of the British sugar cane industry. The same year, Banks sent the Polish gardener Anton Hove to the Maratha territories in northern India, ostensibly to collect for Kew but with secret instructions to obtain cotton plants and knowledge on how to cultivate them and manufacture cotton. At the time, West Indies cotton was inferior to Dutch, Brazilian, and Asian cotton, and the planters wanted to be able to compete better in the market.

ABOVE LEFT *Pierre Poivre, who brought spice plants from the Moluccas to Mauritius.*

BELOW *Calcutta Botanic Garden, depicted here in 1829, was one of many gardens set up in British colonies to cultivate plants of economic value.*

Opuntia stricta

the cactus on which the cochineal insect breeds, from the William Roxburgh Collection held in the archives of the Royal Botanic Gardens, Kew.

Keeping plants ship-shape

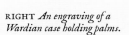

Bringing plants back from foreign climes presented a problem for early collectors. With long sea journeys, many a leafy cargo fell victim to variations in temperature, soaking by sea spray, lack of light, or too little fresh water. In 1829, Nathaniel Bagshaw Ward discovered that plants had a much better survival rate if placed in a closed glazed case with sufficient water and soil. The plants saturated the air with water vapor, which condensed on the glass and kept the soil moist in a continual cycle. The boxes became known as Wardian cases. Hackney nurseryman George Loddiges reported, "Whereas I used formerly to lose nineteen out of twenty of the plants I imported during the voyage, nineteen out of the twenty is now the average of those that survive."

RIGHT *An engraving of a Wardian case holding palms.*

Subsequent Kew directors William Jackson Hooker, Joseph Dalton Hooker, and William Thistleton-Dyer further developed Banks's vision of British imperial botany. By 1889, Kew was shuffling plants between offshoot gardens in Bangalore, Bombay (Mumbai), Calcutta (Kolkata), Madras (Chenai), northern India, Ceylon (Sri Lanka), Mauritius, the Straits settlements (including Singapore, Penang, and Malacca), Hong Kong, New South Wales, Queensland, Tasmania, Victoria, South Australia, New Zealand, Fiji, British Guiana (Guyana), Barbados, Dominica, Trinidad, Grenada, Jamaica, St. Lucia, the Gold Coast (Ghana), Cape Colony (Northern Cape, Western Cape, and Eastern Cape), Niger territories, Natal, Lagos, and Malta. Crops cultivated in alien soils included coffee, oranges, bananas, pineapples, almonds, cochineal cactus, chaulmoogra, ipecacuanha, and mahogany.

The transfer of quinine-containing cinchona from South America to British colonial gardens in India and Ceylon, and Dutch gardens in Java, had a considerable impact on landscapes and livelihoods around the globe. At the time, the only known effective treatment for malaria was an infusion made with bark from the cinchona tree. After the uprising of Indian troops against the British in 1857, the British government was concerned about the health of its soldiers and called on Kew to obtain cinchona plants and develop plantations. One hundred thousand dried seeds and 637 plants obtained by Richard Spruce and Kew gardener Robert Cross in Ecuador were soon growing in the Nilgiri Hills of India and in Ceylon. By 1880, 847 acres (343 hectares) of government plantations and 4,000 acres (1,600 hectares) of private land were growing cinchona in the Nilgiri Hills, and some 5,500 acres (2,225 hectares) of mountainside had been planted with cinchona in Ceylon. It has been suggested that Britain's "home-grown"

LEFT *Richard Spruce helped the British transfer cinchona from South America to India and Ceylon.*

stock of bark helped it to expand into Africa in the late nineteenth and early twentieth centuries. At the time, Africa was the deadliest malarial environment in the world, nicknamed the "white man's grave;" once quinine became more widely available British traders began to penetrate the interior.

The role of plant transfers generally in aiding the expansion of the British Empire is undisputed. That plants and seeds were often acquired without the agreement of the host country, vast swathes of land stripped of natural vegetation to create plantations (causing soil erosion), and millions of people worked as slaves to sustain industries such as sugar cane were not always seen as objectionable issues in the days of Empire. A memorial addressed to Prime Minister William Gladstone in 1873 declares: "The records of the Colonial and India Offices will show of what immense importance the establishment at Kew has been to the welfare of the entire British Empire."

Cinchona pubescens

from *Joseph Rocques'* Plantes usuelles, indigenes et exotiques . . . , 1809.

41

THE TRANSFER OF TEA FROM CHINA TO INDIA

India is today the world's biggest producer and consumer of tea, but before the mid-nineteenth century tea was not grown commercially there at all. The development of the tea trade outside the Orient owes much to a cunning act of bio-piracy executed by the East India Company. In the early nineteenth century, China was the exclusive exporter of tea to the West and restricted trading by demanding payment in silver bullion. Demand for the drink was growing in Europe, so the Company was keen to break China's control of the industry. After tea plants were discovered growing wild in Assam, in northeast India, the British experimented with creating a plantation from both these native plants and seedlings obtained from China. Neither was successful, but in 1848, the Company decided to send a plant hunter to China specifically "for the purpose of obtaining the finest varieties of the Tea-plant, as well as native manufacturers and implements, for the Government plantations in the Himalayas."

The man they sent was Robert Fortune, a Scottish collector who had already visited China between 1843 and 1846 on behalf of the Royal Horticultural Society. During this visit he had gathered plants such as *Platycodon grandiflorus* (which then used the synonym *Campanula grandiflora*), *Abelia chinensis*, *Weigela florida*, and various tree peonies, as well as visiting the tea-producing areas. He had deduced that black tea and green tea were derived from the same plant, and noted that green teas for exporting were dyed with Prussian blue (a slightly toxic substance made by combining prussic acid with iron) and gypsum. This was done on behalf of the foreign "barbarians" from Europe and America who rated the "beautiful bloom" of these teas. The Chinese, meanwhile, drank unadulterated green tea. He had also observed the growing conditions and production methods, and reported that, rather

ABOVE *Tea became a popular drink in America in the early eighteenth century, as this poster advertising The Great American Tea Company shows.*

Camellia sinensis
from the Weyhe Collection in the archives of the Royal Botanic Gardens, Kew.

than being complex or mysterious as people supposed, "the mode of gathering and preparing the leaves of the tea-plants is extremely simple."

In 1848, Fortune set off on his tea-gathering mission. To gain easier access to the interior by blending in with the locals, he shaved his head, leaving just a ponytail, and wore Chinese clothes. He then journeyed for 200 miles (325 kilometers) on foot, by boat, and by sedan chair to the tea-producing district of "Hwuy-chow." Here, Fortune busied himself with collecting seeds, examining

Campanula grandiflora
from Curtis's Botanical Magazine, 1794.

How tea is made

Tea comes from the plant Camellia sinensis, *of which there are two varieties. China tea is made from* Camellia sinensis *var.* sinensis, *while Assam, or Indian, tea comes from* Camellia sinensis *var.* assamica. *The former is hardier and has smaller leaves than the latter. Tea plants can grow to 56 feet (17 meters) high but when grown commercially are pruned to keep them as trim 6.5-feet- (2-meter-) high bushes. Black tea and green tea are made from the same plant, but black tea is fermented during the production process. After picking, leaves have air blown through them to remove moisture and promote enzyme activity. They are then rolled between two surfaces to break the cells in the leaves mechanically and release the juices that give the tea its characteristic flavor.*

Rhododendron fortunei

from Curtis's Botanical Magazine, 1866.

the vegetation of the hills, and obtaining information on the manufacture of green tea. He later visited black-tea-producing areas, including that of "Fokien" in the north. Although there were more accessible tea-producing areas, he wanted to ensure only stock of the best possible quality was transferred to India. In the northern town of "Tsong-gan-hien," he noted, "This city abounds in large tea-hongs, in which the black teas are sorted and packed for the foreign markets . . . Tea merchants from all parts of China where teas are consumed or exported come to this place to make their purchases of tea and the necessary arrangements for its transport."

Fortune despatched his spoils to Calcutta, finding the best method of transporting seeds was simply to plant them in Wardian cases. By the time he wrote up his travels in his 1852 book *A Journey to the Tea Countries of China . . .* , he was able to report that "upwards of twenty thousand tea-plants, eight first-rate manufacturers, and a large supply of implements were procured from the finest tea districts of China, and conveyed in safety to the Himalayas." In time, plantations were established in

Assam and Sikkim, which Fortune astutely predicted would be "likely to prove of great advantage, not only to India but also to England and her wide-spreading colonies." Although China remained central to the tea industry during the second half of the nineteenth century, by 1890 India was supplying 90 percent of Britain's domestic market. Between 1854 and 1929, the value of its tea exports to Britain rose from £24,000 to £20,087,000. Today, tea is the most popular non-alcoholic drink in the world, with more than 3 million tonnes grown each year to meet demand.

Paeonia suffruticosa

from the archives of the
Royal Botanic Gardens, Kew.

SOWING THE SEEDS OF THE WORLD'S RUBBER INDUSTRY

Hevea brasiliensis
from Franz Eugen Köhler's
Medizinal-Pflanzen in
naturgetreuen . . . , 1887–8.

The path of rubber

6th century In Mexico and Central America, Aztecs and Mayans use rubber to coat fabrics and play a game with rubber balls.

1751 French naturalist Charles Marie de la Condamine, with assistance from botanist François Fresneau, publishes the first scientific paper on the properties of natural rubber.

1770 Joseph Priestley coins the name "India-rubber" after seeing caoutchouc being sold in half-inch cubes in London for artists to use to rub out pencil lines.

1823 Charles Macintosh creates a rainproof coat made from two pieces of rubber cloth stuck together with an inner layer of rubber. It becomes known as the "Mackintosh".

1839 Charles Goodyear creates vulcanized rubber.

1853 Traders take advantage of the invention of steamships and penetrate the Amazon in search of rubber supplies.

1888 Henry Ridley conducts large-scale experiments into rubber at Singapore Botanic Gardens and encourages the establishment of plantations. The same year, John Boyd Dunlop invents the pneumatic rubber bicycle tire.

1892 William Tilden synthesizes rubber from synthetic isoprene in the U.K.

1895 Michelin introduces the first pneumatic motorcar tire.

1950s The last wild rubber is exported from Brazil.

1959 The production of synthetic rubber overtakes that of natural rubber.

1990 Around five billion *Hevea brasiliensis* trees are now producing rubber on plantations around the world.

2006 Natural rubber production is 9,680,000 tonnes for the year. Synthetic production is 12,762,000 tonnes.

Natural rubber derives from latex, a white substance produced by many tropical plants. Scoring the bark of these trees—tapping—causes the latex to flow out like a sticky milk. Until the late nineteenth century, the majority of rubber came from Central and South America, where latex-yielding trees grew wild (*Castilla elastica* in Central America and *Hevea brasiliensis* in the rainforest of Amazonia). Some rubber was also found in the wild in India (*Ficus elastica*). The locals had long known of the useful properties of caoutchouc, as they called it. Amazonian Indians used it to waterproof clothing and dwellings against the drenching downpours of the rainforest, while the Aztecs and Mayans used latex from plants such as *Castilla elastica* to create balls for their ritual ball games.

Untreated rubber did not travel well, as it became sticky in hot weather and brittle when cold. This limited its usefulness to the West until 1839, when Charles Goodyear discovered that adding sulfur and lead to rubber and heating it produced a material that stayed dry and flexible even at high and low temperatures. The process became know as "vulcanization" after Vulcan, the god of fire. Soon rubber was being used for everything, from making beds containing warm water to help Welsh miners with hypothermia to elasticating fabrics and insulating undersea cables.

By 1860, the price of rubber had reached an all-time high, equivalent to that of silver. Seeking to avoid relying on South American supplies, the British government hatched a plan for generating its own stocks of the most valuable rubber-yielding species, *Hevea brasiliensis*. The plan was to acquire seeds from South America and set up plantations in colonies with suitable climates.

The government offered to pay one of Joseph Hooker's contacts, Henry Alexander Wickham, £10 per 1,000 seeds for him to obtain

ABOVE LEFT *Henry Wickham, who sent seeds of* Hevea brasiliensis *to Kew.*

BELOW LEFT *Rubber coins from Bolivia.*

BELOW RIGHT *The oldest dated rubber product is this water bottle made in 1814.*

RIGHT *Henry Ridley stands to the left of a rubber tree with herring-bone tapping marks.*

FAR RIGHT *Tapping rubber involves cutting the bark and collecting the latex that flows out.*

seeds of *Hevea Brasiliensis* and send them to Kew. He set out in January 1876. A note in Kew's archives dated 7 July 1876 tells of his success: "70,000 seeds of *Hevea brasiliensis* were received from Mr H. A. Wickham on June 14th. They were all sown the following day, and a few germinated on the fourth day after. Up to this date 2,700 have been potted off—not quite 4 per cent. This may be considered to be the total number of plants, as very few will germinate after this date. Many hundreds are now 15 inches high and all are in vigorous health."

An initial attempt at growing rubber trees in Calcutta had failed because the climate was unsuitable, so Hooker opted to send the plants to Ceylon (now Sri Lanka). He despatched 1,919 *Hevea brasiliensis* and 32 *Castilla elastica* saplings to Dr. G. H. K. Thwaites at the Royal Botanic Gardens, Peradeniya. These were duly planted and were soon thriving. In time, some were transplanted at Henerathgoda Garden, where tapping experiments began. By 1892, one specimen had swelled to a circumference of 77 inches (1.95 meters) and had yielded 7 pounds $2^3/_4$ ounces (3.25 kilograms) of dry rubber in five years.

Hooker also sent plants to Singapore, Jamaica, Montserrat, Queensland, and Cameroon. The 22 seedlings he sent to Singapore Botanic Gardens in 1877 gave rise to a further 1,200 plants from their seeds. These were inherited by Henry Ridley in a somewhat sorry state when he took over as Director of Gardens and Forests in 1888. "I had to clear the rubber field of a dense scrub abounding with snakes including 27 foot [sic] pythons," he wrote. However, his hard work meant he now had a good stock of healthy and accessible rubber. His experiments showed that tapping cuts could be opened every day without harming the tree and would consistently produce the same amount of latex.

In 1907, when a vulcanizing plant opened in Singapore, some of the botanical garden's rubber was used to make the first tires from cultivated rubber. Ridley used these on his dog-cart. Convinced that the demand for rubber from bicycle tires and other such goods would soon outstrip wild supplies, "Mad" Ridley filled the pockets of visiting district officers and planters with seeds to plant around their houses. However, ignorance hindered his cause. "One man came to the office to make enquiries about planting in New Guinea, I suggested rubber," wrote Ridley. "'Not me,' says he, 'I have just heard that they have found a mine of it in America and can dig it out at a penny a pound.'"

By 1930, some 3 million acres (1.2 million hectares) on the Malay peninsula had come under rubber cultivation. The total world production of rubber that year was 821,815 tonnes, most of which came from the Malay peninsula. The next-biggest share, 240,000

Slaves to rubber

King Leopold II of Belgium exploited the Congo's rubber supplies (obtained from a Landolphia vine) by forcing native people to work in the industry. Atrocities there, including the rape and murder of rubber workers, prompted the first human rights movement: Roger Casement and Edmund Morel set up the Congo Reform Association in 1904. Similar atrocities were committed by rubber "overlords" in the Amazon rainforest. In 1912, Casement published the findings of an investigation into the actions of one of them, Julio Cesar Arana. Casement calculated that, in little over a decade, 30,000 natives had been directly murdered or killed by deliberate starvation in the Putumayo region of Peru. Despite a lengthy parliamentary committee investigation of the affair, the British Courts were powerless to imprison Arana and he returned to Peru to continue trading.

tonnes, was from the Dutch East Indies, and Ceylon contributed 62,000 tonnes. Brazil produced only 17,137 tonnes. Today, much of the rubber that is still produced naturally, rather than synthetically, has its roots in the 22 seedlings that were grown from Amazonian stock at Kew and sent to Singapore in 1877.

Castilla elastica
from Franz Eugen Köhler's Medizinal-Pflanzen in naturgetreuen . . ., 1887-8.

A PASSION FOR ORCHIDS

The name orchid derives from the Greek *orchis*, meaning testicle. Theophrastus first used the term in his *Enquiry into Plants*, written around 300 B.C. His choice of nomenclature was a reference to the testical-shaped tubers that certain Mediterranean species exhibit. Orchids were revered in the Far East long before the Greek philosopher first noted them, however. In 2800 B.C., the vibrant pink hyacinth orchid *Bletilla striata* was mentioned by the mythical Chinese Emperor Shennong in *Shénnóng běncǎo jīng*, a text on the medicinal uses of plants. Later, Confucius, who lived between 551 and 479 B.C., called the orchid "the king of fragrant plants." And in the tenth century A.D., Kin-shō's *Orchid Book* gave a history of oriental cymbidiums, with the names of the first growers, geographical locations, and growing techniques.

One of the earliest Europeans to write of orchids was Engelbert Kaempfer, a German naturalist and physician working for the Dutch East India Company (VOC). After 1639, the VOC had exclusive trading rights with Japan, and Kaempfer visited several times. In his account *Amoenitatum exoticarum . . . Fasciculi* (1712), he mentions *Dendrobium moniliforme*. This plant was grown by friends of the imperial aristocracy of Kyoto, who named it *sekkoku*, meaning "an orchid that makes men live a long life." By 1698, the first tropical orchid to grow in Europe, from Curaçao, was thriving in Holland. Within decades, a dried specimen of *Bletia purpurea* had sprung into life on a bed of bark in the London garden of Sir Charles Wager, a Royal Navy officer.

Kew Gardens received its first tropical orchid, *Epidendrum rigidum*, in 1760. The speed at which Kew's collection subsequently

Orchid-hunting personalities

Orchid hunters have come from a wide variety of backgrounds. Some of the most prominent included:

JÓZEF RITTER VON RAWICZ WARSZEWICZ *(Born Poland, 1812–66): He fled Poland after the failed Polish Revolution. When seeking asylum in Guatemala, he encountered orchids and dedicated his life to collecting and sending specimens back to Germany. He is remembered in:* Cattleya warscewiczii.

JEAN LINDEN *(Born Luxembourg, 1817–98): He gathered some 1,100 species from Brazil, Mexico, Cuba, Central America, the Andes of Colombia, and Venezuela. He is remembered in the ghost or white frog orchid:* Dendrophylax lindenii.

REVEREND CHARLES SAMUEL POLLOCK PARISH *(Born England, 1822–97): He worked as a missionary in Burma and grew 150 species in the mission garden. He is remembered in:* Paphiopedilum parishii.

BENEDIKT ROEZL *(Born Prague, 1823–85 [right]): He traveled extensively through Latin America, sending back 800 species to Europe. He is remembered in:* Miltoniopsis roezlii, syn Odontoglossum roezlii.

Odontoglossum rossi amesianum
from Robert Warner and Benjamin Samuel Williams's The Orchid Album, 1882–97.

46

expanded reflects the growing interest in orchids at the time. By 1768, the gardens boasted 24 species, two of which were tropical, the rest indigenous. By 1789, 15 exotic species were in cultivation; and by 1813, the gardens contained 46 tropical species, with about 12 more from Australia and South Africa. Then, in 1818, William Swainson collected some orchids from Brazil and sent them back to England. When one flowered later that year at the home of the recipient, William Cattley, its beautiful, huge, trumpet-like labellum caused a sensation. Cattley described it as "the most splendid, perhaps, of all *Orchidaceous* plants . . . ." It was named *Cattleya labiata* in his honor.

Swainson did not publicize the location from which he had taken the plants, but in 1836, the naturalist Dr. George Gardner claimed to have found some growing in the Organ Mountains of Rio de Janeiro. It later transpired that Gardner had not found *Cattleya labiata* at all; he had spotted *Sophronitis lobata*. Gardner also claimed to have found the elusive *Cattleya labiata* on the banks of the Parahyba river but this turned out to be *Cattleya warneri*. It was

East. Competition between collectors was intense, as this quote from Theodor Cordua about collector Carl Hartweg Theodore demonstrates: "I have often heard him say that Mr. George Ure Skinner [an orchid collector] and himself discovered the large plant of *Laelia superbiens* both at the same time when in Mexico, and that they were both determined to have it, but could not get it then, for it was up a very high tree. Hartweg outwitted Mr. Skinner by going early in the morning, taking with him a native and an axe, and chopping the tree down, at the same time conveying away the large *Laelia.*"

not until 1889 that *Cattleya labiata* was rediscovered in Pernambuco, the location from which Swainson had originally taken the first specimens. *The Orchid Review* hailed it as "the event of the year."

Another orchid, or orchid product, that excited the attention of Europeans was vanilla, extracted from the aromatic pods of *Vanilla planifolia* or *Vanilla fragrans*. Although by the late seventeenth century it was quite well known across Europe, the Spanish still controlled supplies. The French and English were keen to cultivate vanilla plants for themselves rather than rely on the Spanish, but they had no idea how to cultivate the plant outside its native habitat. The breakthrough came only after the French took plants to the Indian Ocean island of Réunion. A slave boy named Edmond Albius managed to pollinate a vanilla vine by peeling back the lip of a small orchid flower with his thumb, lifting the rostellum out of the way and pressing the anther and stigmatic surfaces together. After this discovery, numerous plantations on Réunion began growing vanilla. Within 50 years, Réunion was exporting 200 tons annually, outstripping Mexico as the world's largest producer of vanilla beans.

By now, nurseries had joined the ranks of those keen to obtain new plants from far-flung habitats. The Nurseries of James Veitch (later James Veitch & Sons) were the first commercial nurseries to send their own plant collector abroad, dispatching William Lobb to South and North America and Thomas Lobb to the Far

Coaxing vanilla into cultivation

The main orchid that yields the aromatic pods is Vanilla planifolia. *It is a vine with small, yellow, almost scentless flowers. When pollinated, the ovaries swell to form long green beans that contain thousands of tiny black seeds. When these pods turn brown after a long drying and curing process, they emit the aroma we now call vanilla. In nature, vanilla orchids can only be fertilized by a tropical bee native to Central America. The seeds need a symbiotic relationship with a fungus in order to germinate. They are eaten by bats in the wild and then grow wherever the seeds are excreted if the conditions are favorable.*

A NEW GENRE OF ART

Convolvulus vulgaris major albus

painted in watercolor by
Georg Dionysius Ehret (1708–70).

Although animals featured in early cave paintings, plants did not appear in artistic works until much later. Stone bas-reliefs created in the fifteenth century B.C. in the Great Temple at Karnak, Egypt, depict 275 plants, including identifiable pomegranates and the dragon arum *Dracunculus vulgaris*. There are also recognizable images of the Madonna lily (*Lilium candidum*) on a bronze-age Minoan wall painting at Amnisos near Knossos, Crete.

It was at this time that the Greek physician Dioscorides wrote *de Materia Medica*, five volumes on the preparation, properties, and testing of drugs. The earliest surviving illustrated botanical work is a sixth-century copy of his work called *Codex Vindobonensis*. Its rather stiff and unrealistic images set the style for botanical illustration until late in the fifteenth century, when artists began taking inspiration directly from nature. More accurately portrayed plants appear in Brunfel's 1530 *Herbarum Vivae Eicones* and Leonhart Fuch's 1542 *De Historia Stirpium*. Leonardo da Vinci adopted this new approach, writing in his *Codex Atlanticus* that he had produced "many flowers portrayed from nature."

In the seventeenth century, natural science flourished as explorers encountered exotic species of plants and animals in new parts of the globe. The artist Maria Sibylla Merian became fascinated by the process of metamorphosis, and published three volumes of illustrations depicting European insects at their different life stages on their favored food plants. She later spent two years in Surinam, after which she published *Metamorphosis Insectorum Surinamensium*. Not everyone needed to travel far to see foreign plants, however, as new varieties were flooding into Europe. As wealthy individuals and botanic gardens built up collections of unusual plants, they began commissioning artists to illustrate them in sumptuous books called florilegia.

During the mid-eighteenth to mid-nineteenth centuries, botanic illustration evolved further to meet the needs of science. As Linnaeus based his classification on the reproductive organs of plants, so artists began emphasizing these parts to aid identification. Georg Dionysius Ehret was one exponent of Linnaeus's system and went on to become the dominant botanical illustrator of this golden age. The son of a gardener, Ehret was born in Heidelberg, in Germany. After settling in England in 1736, he instructed high society in

LEFT *This bronze-age Minoan wall painting at Knossos clearly shows the Madonna lily.*

Codex Vindobonensis

Codex Vindobonensis Med. Gr. 1., *the earliest surviving illustrated botanical work, is a copy of Dioscorides's* De Materia Medica *made in the year 512 for Juliana Anicia, daughter of the former Western Roman emperor Olybrius. Made by copying earlier works, it was seen by Ogier Ghiselin de Busbecq, the ambassador from the court of Ferdinand I to Suleiman the Magnificent, in 1562. Seven years later, the manuscript found its way into the Imperial Library at Vienna, but no one knows whether the purchaser was Busbecq or the Emperor Ferdinand. After World War I, the codex was seized by the Italians and removed to Venice, but was later returned to Austria. The book contains nearly 400 full-page paintings of plants. The styles of some suggest they were derived from illustrations dating back to the second century A.D.*

botanical art and contributed to florilegia and travel books. These included his own *Plantae et Papiliones Rariores*, Trew's *Hortus Nitidissimis*, and Hughes's *The Natural History of Barbados*.

After Ehret died, John Miller made plates for the 1777 *Illustrations of the Sexual System of Linnaeus*, which Linnaeus described as "more beautiful and more accurate than any that had been seen since the world began."

Sir Joseph Banks was to play a role in directing botanical art in the late eighteenth and early nineteenth centuries. As well as commissioning Sydney Parkinson to accompany him on Cook's first circumnavigation, he employed Franz (also known as Francis) Bauer to work at Kew Gardens as "Botanick Painter to His Majesty." Bauer made detailed plant drawings, often at microscopic level, which were published in volumes such as *Delineations of Exotick Plants* (1796–1803) and *Strelitzia Depicta* (1818).

Franz's brother Ferdinand was also a botanical artist. He traveled to Australia, where he made highly detailed studies of plants, even illustrating the intricacies of their pollen grains. He was also responsible for producing the first accurate illustrations of Australian wildlife.

From 1787, William Curtis's *The Botanical Magazine* offered less wealthy people a chance to own beautiful images of flowers and plants. It was published monthly throughout Queen Victoria's reign, with each issue containing 60 hand-colored plant portraits, mostly of new introductions.

Its finest illustrator was Walter Hood Fitch, a Scotsman, who contributed over 2,000 illustrations between 1837 and 1878. He was employed by William Jackson Hooker, who edited the magazine and was also Director of Kew from 1841 to 1865. Fitch also contributed work for Hooker and George Bentham's *Handbook of the British Flora* (1865), plus Joseph Hooker's *The Rhododendrons of Sikkim-Himalaya* (1849–51) and *Illustrations of Himalayan Plants* (1855). Still published today, *Curtis's Botanical Magazine* is now the longest-running botanical periodical that features color illustrations of plants.

Nigella hispanica
from Curtis's
Botanical Magazine, 1810.

CONSERVING PRECIOUS PLANTS

As far back as 1799, Baron Alexander von Humboldt expressed concern at the rate of felling of cinchona trees in South America. Following in his footsteps, 50 years later, Richard Spruce concluded that wild supplies of plants used as commodities would run out unless cultivated. And when George Gardner arrived in Ceylon in 1844 to take charge of the Royal Botanic Gardens, Peradeniya, he noted, "Botanists of future time will look in vain for many of the species which their predecessors had recorded in the annals of science as natives of the island."

Evidence that humans were having a detrimental effect on the environment was also being recorded in the U.S. In 1847, George Perkins Marsh gave a speech drawing attention to the destruction of forests and calling for an approach to their management based on conservation. Over the next half-century, Yellowstone, Sequoia, Yosemite, and General Grant National Parks were created with the aim of preserving wilderness areas. Meanwhile, several campaigns in the U.S. and Europe began highlighting environmental problems of disease and air and water pollution.

After World War II, concern about the impact humans were having on the environment became more widespread. In response, the world's first global environmental organization, the International Union for Conservation of Nature (IUCN) was set up in 1948. Then in 1970, the first-ever list of threatened plants was published, entitled the *Red Data Book*. Its author, Ronald Melville, drew the startling conclusion that 20,000 plant species were in need of some form of protection to ensure their safety and continued existence.

The scientific world soon drew the conclusion that certain animals and plants were under threat through commercial exploitation. As a result, the Convention on International Trade in Endangered Species of Wild Fauna and Flora (CITES) came into force in 1975. To date, 175 countries have signed up to the convention. There are three levels of protection. Appendix I includes species that are the most threatened among CITES-listed animals and plants. International trade in wild specimens for commerical purposes is banned for these species. Appendix II lists species that are not necessarily now threatened with extinction but may become so unless trade is closely regulated by export permit. Species are listed under Appendix III to ensure assistance from other CITES parties in regulating the trade. Around 30,000 species of plants are now protected by CITES from overexploitation through international trade.

A memorial to threatened plants

The artist Willem Boshoff has built a memorial garden to 15,000 rare plant species that could disappear very soon. The installation contains 15,000 cloth flowers, each printed with the details of a threatened plant. These have been "planted" in the main lawn at Kirstenbosch Botanical Gardens in South Africa. The installation resembles the graveyards commemorating soldiers who died in World War I; Willem was inspired after seeing the fields of poppies at Ypres in Flanders. The purpose of the memorial garden is to illustrate the loss to the world should these presently threatened plant species become extinct. South Africa has the highest concentration of threatened plants in the world. Of the country's 22,102 plant species, more than 1,500 face a high risk of extinction in the near future.

BELOW *Yellowstone National Park was founded in 1872 to help preserve wilderness areas.*

RIGHT Bertholletia excelsa *from
Thomas Croxen Archer's* Profitable
Plants, *1865. Yields of the nut depend
on the health of the rainforest.*

BELOW *The Amazon rainforest and
river system contain a fifth of the
world's fresh water and one in ten of
the Earth's species.*

In 1992, the first international Earth Summit convened in
Rio de Janeiro, Brazil, to address the urgent problems of
environmental protection and also socio-economic development.
The heads of state in attendance signed documents outlining
policies for achieving sustainable development to meet the
needs of the poor and to recognize the limits of development to
meet global needs. Among the treaties was the Convention on
Biological Diversity (CBD), which aims to conserve biodiversity,
promote the sustainable use of plants and animals, and ensure that
the benefits arising from genetic resources are shared out equally.
As of 2011, however, 9,107 of the world's 313,655 described species of
plants, mushrooms, and lichens were considered by the IUCN to
be threatened with extinction.

As botanical knowledge developed through the 1970s and 1980s,
scientists became more aware of the importance of saving entire
habitats, rather than individual species. Sir Ghillean Prance, who
was at the New York Botanical Garden between 1963 and 1988 and
was Director of Kew Gardens between 1988 and 1999, conducted
pioneering research into the interdependencies between plants
and animals. After many expeditions to the Brazilian Amazon,
he concluded that a successful wild harvest of the commercially
valuable brazil nut was dependent on the health of the
surrounding rainforest. This is because the tree requires female
euglossine bees to pollinate it, and they will only mate with males
who successfully gather a cocktail of scents from several orchid
species, all of which grow only in undisturbed forest.

The botanical artist Margaret Mee was also a regular visitor
to the Amazon between the 1950s and 1980s. After witnessing the
destruction of the rainforest first-hand, she began including the
natural habitat of the flowers she painted in order to emphasize
the interdependency between plants and their environment. In
1988, shortly before her death, she wrote: " . . . the [Amazon]
forest has changed considerably and the lovely plants I have
painted along the Rio Negro have disappeared. I can remember
my excitement on my first journey there, when I moored my boat
to a Swartzia tree, full of perfumed white flowers, among the
great trees on the banks. The changes have been disastrous, and
the destruction and burning of the forests arouse fears for the
future of our planet."

Sir Ghillean Prance (1937–)

*Sir Ghillean Prance's boyhood pastime of collecting
flowers, birds' eggs, and insect larvae soon developed
into a passion for botany. He studied the subject at
Oxford before accepting an invitation from the New
York Botanical Garden (NYBG) to join an expedition
to Surinam on the northern edge of the Amazon basin.
Over the next decade, he spent a great deal of time
exploring the rainforests' plants; his achievements
include identifying a new tree species,* Acioa edulis.
*After realizing the damaging impact the 3,400-mile
(5,500-kilometer) trans-Amazonian
Highway would have on the forest, he
shifted his emphasis to conservation.
He spent 25 years rising through the
ranks at NYBG, before accepting the
post of director at Kew in 1988. He
inspired Kew to focus on conservation
and the sustainable use of plants.*

THE ROLE OF THE MODERN BOTANIC GARDEN

The role of botanic gardens has changed down the centuries. The first gardens of the Italian Renaissance were set up to grow medicinal plants and help physicians learn more about herbal remedies. The grand gardens of the eighteenth century were designed to showcase exotics, with glasshouses such as the Palm House at Kew built to re-create tropical habitats being encountered by navigators and explorers. Botanic gardens developed in the colonies during the nineteenth century became experimental scientific stations to which plants from similar habitats in other countries could be transferred and grown for the commercial gain of European nations. "Gardens reflect the era that they're created in and the prevalent thinking of the time," says Sara Oldfield, Secretary General of Botanic Gardens Conservation International (BGCI).

BGCI was set up in 1987 to assist botanic gardens in using their resources and knowledge to help conserve plants threatened in the wild. Today, from its headquarters beside Kew Gardens in London, it coordinates the conservation efforts of the 2,500 botanic gardens in 120 countries around the world. One of its roles has been to help develop the Global Strategy for Plant Conservation, an action plan for plants that grew out of the Convention for Biological Diversity. Over 180 countries backed the strategy in 2002. Recognizing that up to two-thirds of the world's plant species could be threatened by the end of this century unless urgent steps are taken, it lays down 16 targets to be achieved by 2010. These include compiling a list of known plant species as a step toward creating a world flora; conserving at least ten percent of the world's ecological regions; and ensuring that 30 percent of plant-based products come from sustainable sources.

At Kew Gardens, specialized techniques of micropropagation are helping it bring back threatened

FAR LEFT The lady's slipper orchid, Cypripedium calceolus.

LEFT A new botanic garden is being created in Amman, in Jordan, to help preserve wild species of agricultural crops.

plant species from the brink of extinction. It helped Australian scientists conserve the wollemi pine. One British success story is that of the lady's slipper orchid (*Cypripedium calceolus*). Over time, the species had been reduced to a single plant in the wild, through overzealous collecting by Victorian orchid enthusiasts. Attracted by the plant's yellow pouch and deep red sepals, they had gathered plants in their thousands. However, the collectors were mostly unsuccessful at growing them because the orchid thrives only in open chalk grasslands or pine woodlands. Initially, Kew's specialists were unable to germinate seeds of the orchid, but eventually an expert in Canada got in touch to say that he had managed to germinate another *Cypripedium* species by using immature seeds.

The Wollemi pine (*Wollemia nobilis*)

In 1994, officers working for the New South Wales National Parks and Wildlife Service discovered a stand of unusual pine trees growing in the Blue Mountains 95 miles (150 kilometers) from Sydney, Australia. They were found to belong to a new genus, related to the monkey puzzle and kauri pine. Fossil records of the family Araucariaceae, *to which the three species belong, date back to the time of the dinosaurs. Kew took part in a propagation program to prevent the 100 or so remaining trees from being targeted by plant collectors. You can now buy a Wollemi pine from outlets around the world, so there is no need for anyone to source one from the wild.*

LEFT Kew's Palm House was built to house tropical plants that were brought to London by its collectors from far-off locations.

Another new botanic garden being set up on a 450-acre (180-hectare) site near Amman, Jordan, also aims to preserve its native plants. This is of global importance, because many of the wild relatives of food crops, such as wheat, barley, oats, garlic, onion, lentils, pistachios, almonds, and apricots, come from the region. It was these dryland plants that enabled civilization to evolve and spread out from the Middle East. However, centuries of cultivation mean that many of the plants grown as staple foods are genetically less diverse than their wild ancestors. As a result, they can be nutritionally inferior, susceptible to disease, and less able to adapt to climatic changes. As climate change takes hold, knowing how to preserve these wild food stocks may prove vital to sustaining current human populations.

Meanwhile, a Swedish orchid grower and pediatrician had found that growing them in a cocktail of nutrients usually fed to premature babies could also coax orchid seeds into life. When Kew's experts tried growing immature seeds in jelly containing the concoction, they began to get almost 100 percent germination rates. Staff in the Micropropagation Unit now grow several hundred of the orchids each year, and have supplemented the lone surviving wild example with around 100 more plants. The hope is that other rare U.K. orchid species may now also be saved using the same techniques.

Kew is also using its knowledge and resources to help other botanic gardens carry out conservation projects. In 1995, the Soufrière Hills volcano on the Caribbean island of Montserrat rumbled to life for the first time in 400 years. Over the next two years, it unleashed ash, gas, and mud flows, burying the capital city of Plymouth and several villages. Among its casualties were the island's botanic garden and wide areas of formerly pristine cloud forest inhabited by bats, lizards, butterflies, and birds. Today, over half the island's vegetation is dead but the cloud forest remains relatively intact in the Centre Hills area. Kew has been helping the islanders to assess the biodiversity of this area, create a vegetation map and "Red List" of threatened species, and set up a new botanic garden to showcase the island's native flora.

Of the island's three endemic species, two have been saved by cultivating them in the new botanic garden's nursery and stockpiling their seeds in the Millennium Seed Bank at Wakehurst Place in West Sussex, England. These are *Rondeletia buxifolia*, a member of the coffee family, and the orchid *Epidendrum montserratense*. The third endemic, *Xylosma serrata*, has not been seen since 1979, although dried specimens are held at the Natural History Museum in London.

ABOVE *Volcanic eruptions in Montserrat swallowed up the capital, Plymouth, along with the original botanic garden.*

MODERN-DAY PLANT HUNTERS

Although methods and motivations may have changed since the collecting trips of earlier centuries, the days of the plant hunter are far from over. One of the targets of the Global Strategy for Plant Conservation is to compile a list of all known plant species as a step toward creating a world flora. It is an ambitious goal that is still a considerable way from being achieved, not least because there are still many species out there to find, name, and describe. At present, 2,000 species of plants new to science are discovered every year. In the past decade, scientists have discovered five new species of palm and a rhododendron with a giant flower in a "lost world" in New Guinea; 50 new species and varieties of plants and fungi in Cameroon; and a huge new flowering palm (*Tahina spectabilis*) in Madagascar.

Western Australia is proving a particularly rich hunting ground for new species. Greg Keighery has spent 34 years studying the flora of the region, which is Australia's largest state and covers around a third of the continent. During his career, Greg has described 75 new taxa of flowering plants and two new genera of flowering plants, as well as adding 32,000 specimens to the Western Australia Herbarium. He currently spends around three months of the year on plant-collecting trips. The fruits of his searches include *Stylidium keigheryi*, a low-growing, pink-flowered triggerplant, and *Eleocharis keigheryi*, a sedge that produces green inflorescences in spring. "Western Australia is an area of mega diversity of flowering plants and we urgently need basic information about our unique species," he says.

BELOW *Wakehurst Place's tree collection was improved following the hurricane of 1987.*

In Wilson's footsteps

While gathering seeds in China, Tony Kirkham and his long-term traveling partner, Mark Flanagan, became interested in his plant-hunting predecessor, Ernest Wilson. They found that Wilson had often visited places before him, and when they started looking at herbarium specimens to make collecting lists, Wilson's name kept cropping up as having supplied the specimens. "In 2003, we went to a place called Moxi in China," recalled Tony. "We were having dinner with the mayor and asked if there were any really big trees in Moxi. He said yes, there was one, but that it was dead. When we went to see it, we realized we'd seen it before in a book, so we sent an email to Kew and asked them to find the photograph, scan it, and send it to us. In the morning it came back and it was the same tree, taken by Ernest Wilson in 1908."

BELOW *Tony Kirkham, Kew Gardens' resident tree expert.*

BELOW RIGHT *Scientists only recently discovered this flowering palm,* Tahina spectabilis, *in Madagascar.*

Not all modern-day plant-collecting trips are designed purely to extend our botanical knowledge, however. Plant-collecting husband-and-wife team Bleddyn and Sue Wynn-Jones have traveled as far afield as Jordan, Taiwan, Japan, Nepal, Sri Lanka, Vietnam, Guatemala, and Korea in their effort to track down new plants for their specialist nursery, Crûg Farm Plants, in north Wales. As well as earning their living by selling the unusual specimens they bring back, they aim to conserve plants by getting them to grow in as many different places as possible. Plants they have introduced to horticulture in the U.K. include *Clematis szuyuanensis* and *Actaea taiwanensis* from Taiwan. "We go with the aim of finding something new, something that's not in cultivation," says Bleddyn.

Tony Kirkham, meanwhile, was lured into plant hunting by the great storm that struck the United Kingdom in 1987. As head of the Arboretum at Kew Gardens, he was responsible for the garden's hardy woody plant collection, which had been badly depleted by the hurricane. Instead of simply replacing lost species, Tony and other staff used the storm as an opportunity to improve the collections at both Kew and Kew's sister garden, Wakehurst Place, in West Sussex. They looked at where there were taxonomic weaknesses in

Kew's collection and geographic weaknesses in Wakehurst Place's collection and set out to find and bring back seeds of the required plants. "Between 1989 and 2003 we visited countries around what we called the 'temperate loop', a circle incorporating South Korea, Taiwan, Hokkaido, Russia's Far East, plus China," explains Tony.

Tony collects only seeds when he goes on plant-collecting trips. This is because young seedlings of trees do not acclimatize well to the new conditions of a different country. Also, plants grown from seed are sexually propagated, so they contain a good range of genetic material. It is rather like with human beings; children of the same parents have different combinations of genes. Every time Tony collects seeds, he takes a fertile specimen of leaves and flowers from the tree from which they came. This is dried and goes into Kew's Herbarium. "The herbarium specimen can be more valuable than the living plant," he says. "We might give something the wrong name in the field because it's not flowering or it's a poor specimen, but if we give the dried specimen to a taxonomist when we get back, they'll tell us the correct name. The living plant might die or might never get established but the herbarium specimen will be there forever."

RIGHT Stylidium keigheryi, *which is named after the botanist who discovered it, Greg Keighery.*

BELOW *A specimen sheet from the Herbarium at the Royal Botanic Gardens, Kew. Each plant for which specimens are collected may have several sheets to shows samples of seed pods, flowers, leaves, and, if appropriate, the bark.*

LEFT *Greg Keighery, who works for the Department of the Environment and Conservation, Government of Western Australia.*

BELOW *Several recent plant discoveries have been made in New Guinea's "lost world".*

INVADERS OF THE PLANT WORLD

One unwelcome side effect of the myriad transfers of plants and seeds around the world is the translocation of "invasive" species. Plants arriving on foreign shores with an agreeable environment and a lack of predators have often quickly become naturalized. Those also encountering a ready pollinator or suitable means for dispersing seeds have been able to spread rapidly. In some cases, the new conditions have made the plant much more successful in its new locale than in its indigenous habitat. When a plant becomes disruptive to native flora in a particular location, it is deemed invasive. For example, *Clematis vitalba* is an innocuous climber in its homelands of south, west, and central Europe. However, since becoming naturalized in New Zealand in the 1930s, it has rampaged through scrubland and forests, smothering 65-foot (20-meter) trees in its path.

Plant hunters, and the botanic gardens or nurseries that employed them, have played a major role in introducing alien plants. For example, the Australian mock orange *Pittosporum undulatum* was spread initially via the network of British colonial botanic gardens. It was introduced to the Cinchona Botanical Gardens in Jamaica in 1870 and is now invading relatively undisturbed montane rainforest. The plant flowers and sets fruit earlier than most of the island's native trees, so there is little competition for pollinators. Its foliage contains oils and resins, which may poison nearby plants, and it produces many seeds, which are spread by birds into new areas. All these factors have made it highly successful in new territories. As well as invading

Pittosporum undulatum

from Henri-Louis Duhamel's
Traité des arbres et arbustes . . . , 1800-19.

Jamaica, it has become problematic in South Africa, parts of Australia that are outside its natural range, and Hawaii.

The brightly colored flowers of *Lantana camara* made it a popular garden flower in Europe when it arrived there from Central and South America. As the colonial powers expanded into the tropics it, too, became widely dispersed. Today, it is considered a problem in at least 50 countries. Since it was introduced to South Africa in 1880, it has invaded natural forests, plantations, overgrazed or burnt veld (grassland), orchards, rocky hillsides, and fields. It arrived on Floreana Island in the Galapagos Islands in 1938 as an ornamental. Since 1970, it has replaced *Scalesia pedunculata* forest and dry vegetation of *Croton*,

Clematis vitalba

from Friedrich Gottlob Hayne's
Getreue darstellung und
beschreibung . . . , 1805.

ABOVE *Lantana camara looks colorful but is an invasive weed that spreads quickly when birds disperse its seeds.*

Macraea, and *Darwiniothamnus.* Two of the three populations of *Lecocarpus pinnatifidus* and one of *Scalesia villosa,* both endemic to Floreana, the smallest island in the Galapagos, face elimination if the invader continues to advance. If *Lantana* reaches the crater area of Cerro Pajas, it will endanger the last remaining nesting colony of dark-rumped petrels on the Galapagos Islands. Thorny thickets of *Lantana* are so dense they would prevent the birds from making their nesting burrows at the breeding site.

Rhododendron ponticum was also introduced as an ornamental. Originating in southeastern Europe and western Asia, it arrived at Kew in 1793 and was soon being distributed by nurseryman Charles Loddiges. Wealthy Victorian aristocrats used it as cover for pheasants on their country estates, from where it escaped to invade semi-natural woodlands on acidic soils. Members of the genus *Rhododendron* are rarely invasive, but this one is an exception. It moves rapidly into mixed oak woodlands, where it shades out bryophyte, herbaceous, and dwarf shrub layers. A single *Rhododendron ponticum* plant, with its branches rooting where they touch the soil, can cover 1,075 square feet (100 square meters). On the British island of Lundy, *Rhododendron ponticum* is threatening the Lundy cabbage (*Coincya wrightii*) and the flea beetle (*Psylliodes luridipennis*) that feeds on it. Clearing the invader from here is difficult, as it grows on steep cliffs and in gullies. In 1997, it took 226 volunteer-hours to clear 2.5 acres (1 hectare).

Five of the worst culprits

***Imperata cylindrica*: Cogongrass**
This native of southeast Asia, which is reported as invasive in 73 countries, infests pine woodlands, dunes, wetlands, and grasslands. Gardeners often grow a red-tipped form called Japanese blood grass that appears to be less invasive than the wild, green type. However, scientists are concerned that the two could hybridize, making a hardier version of the invader that would thrive in colder areas.

***Fallopia japonica*: Japanese knotweed**
This has become a major pest in the U.K., Europe, Australia, New Zealand, and North America. It forms dense stands that, when disturbed, send out underground stems (rhizomes) several feet long. This enables it to start a new population away from the disturbance. This canny ability makes it very hard to get rid of, even with chemicals.

***Eichhornia crassipes*: Water hyacinth**
Despite being considered one of the worst weeds in the world, this Brazilian native is often grown as a garden ornamental in ponds. It has large mauve flowers that sit atop floating spongy bladders. It thrives in all freshwater environments, from lakes to ditches, and grows in such profusion that it can stop boat traffic.

***Hiptage benghalensis*: Hiptage**
The attractive, fragrant flowers of this plant belie its murderous habits. A climber, it grows up into tree canopies then smothers the native vegetation. It has been reported as invading La Réunion, Mauritius, Hawaii, and Florida. The IUCN considers it one of the worst 100 invasive organisms in the world.

***Heracleum mantegazzianum*: Giant hogweed**
This plant's huge leaves and 10-foot- (3-meter-) high flower heads make it stand out from the crowd. It seeds freely and forms a canopy that displaces native plants in forests and along waterways. Invasive in the U.S. and the U.K., it contains chemicals that can burn human skin, leaving it permanently scarred.

STOCKPILING SEEDS FOR POSTERITY

In the mid-1990s, scientists at Kew Gardens, London, came up with the idea of creating a global seed bank specifically for conservation uses. By enabling new populations to be established, the bank would act as a lifeline for plants threatened by deforestation or desertification. In 1995, Kew successfully submitted a proposal to the Millennium Commission for the Millennium Seed Bank Project, to be located at its sister garden, Wakehurst Place, in West Sussex. The Wellcome Trust Millennium Building, designed especially to house the seed collection, opened in 2000. In 2008, it contained seeds of 96 percent of the U.K.'s flora, and by 2010 it had collected seeds from more than 10 percent of the world's flora. The intention is for it to contain seeds from a quarter of the world's flora by 2020. "So far we have seeds from 30,855 species, which amounts to more than one and a half billion seeds," says Paul Smith, Head of the Millennium Seed Bank. "The Millennium Ecosystem Assessment reckons that between 60,000 and 100,000 species are threatened with extinction, around one third of the total. We have enough space here for about half the world's species, so we can certainly house all of the world's rare and threatened species."

The Millennium Seed Bank is already demonstrating its value. For example, it holds seeds of several species that are either

ABOVE *The Millennium Seed Bank at Wakehurst Place aims to house seeds of ten percent of the world's flora by 2010.*

BELOW *Storing seeds can act as an insurance policy to help ensure future supplies of important drought-tolerant plants, such as* Hyphaene thebaica *(foreground) and food crops, such as the date palm,* Phoenix dactylifera *(background).*

Centaurea cyanus
from Friedrich Gottlob Hayne's
Getreue darstellung und
beschreibung . . . , 1805.

Excavating ancient seeds

Efforts to conserve seeds have not been confined to those occurring in the wild, but have even extended to archaeological investigations. Dr. Fiona Hay, Collection Studies Team Manager at the Millennium Seed Bank, has visited Peru to see if seeds buried with ancient mummies could be coaxed into life. Inca farmers are known to have cultivated some 500 types of plants in the fifteenth century, spreading nutrient-rich anchovies on the land as fertilizer and building sophisticated systems of terraces and canals to irrigate their seedlings. Some of the seeds buried with mummies date back even further. "Peru is so dry that plant remains are well preserved," says Dr. Hay. "Whole maize cobs and lots of cotton had been buried with some of the mummies."

RIGHT *The Svalbard International Seed Vault, in Norway, which aims to safeguard food crops, has been built deep into the side of a mountain in the Arctic.*

extinct in the wild or extremely rare. U.K. natives include *Bromus interruptus* (interrupted brome), a grass that has been extinct in the wild since 1972, and *Centaurea cyanus* (cornflower), which is threatened by modern agricultural methods. Foreign species include *Cylindrophyllum hallii*, a native of South Africa that is one of the rarest plants in the world. The species had not been seen since 1960 when a group of collectors went to gather seeds from it in 2001. After much searching, they found a population of it in the Northern Cape. However, over half of the 300 plants were either dead or dying. The team collected seeds from 85 plants, taking less than five percent of what was available. Back at the Millennium

Seed Bank, these gave rise to 107 plants, which are now yielding new seed stocks. A return visit to the collection site in 2002 revealed that all the original plants had died; only six plants remained in a separate population nearby. The stored seed means South African conservation organizations can reintroduce the plant in future if they choose to do so.

The Millennium Seed Bank is very much a working bank. Its seeds have been used in projects all over the world. More than 150 collections of salt-tolerant pasture species from the collection have been used in trials in Australia, where some 14.1 million acres (5.7 million hectares) are affected by salination. Seeds of drought-tolerant species are meanwhile helping to boost agricultural productivity and combat desertification in Egypt. And collections of seeds from medicinal plants are helping Pakistan produce remedies from cultivated plants instead of wild resources. The nation requires 20,000 tons of plant material for medicines each year.

Each time that seeds are collected, the scientists record the exact location, along with details of the number of plants in each population and any threats facing them. By the end of 2009, the Seed Bank will have gathered data on circa 55,000 plant populations around the world. As climate change takes hold, this data and future observations will be invaluable in showing how different species are responding to new temperatures and weather regimes, and help in pinpointing species that should be conservation priorities.

BELOW *Gathering seeds in the field. Scientists record the exact location of specimens taken, along with information on plant populations.*

Bromus interruptus

from the London Journal of Botany, 1842-57.

The Doomsday Vault

A recent addition to the global network of seed banks is the Svalbard International Seed Vault, located deep in a remote mountainside close to the North Pole. It aims to safeguard all known varieties of food crops. Dubbed the "Doomsday Vault," it will act as a backup store for the global network of seed banks. In choosing its location, experts assessed the likely impacts of climate change and natural catastrophes. They chose a site that would not be affected by floodwaters and which would be cold enough to provide natural refrigeration in the event of a power breakdown. The fate of the Philippines' national seed bank exemplifies why backups are vital: it was destroyed in 2006 when a typhoon ripped through the region.

THE EFFECTS OF CLIMATE CHANGE

The world is 1.37°F (0.76°C) warmer today than it was in the latter half of the twentieth century. Apart from a leveling out of the temperature between the mid-1940s and mid-1970s, the planet has been steadily warming up since 1910. This is because, since the industrial revolution, ever-increasing numbers of factories, refineries, power stations, cars, and aeroplanes have been spewing gases into the atmosphere. These have enhanced the natural greenhouse effect, which is the process by which Earth regulates its temperature. Scientists predict the temperature could rise by as much as 7.2°F (4°C) by the end of the century.

If there were no greenhouse effect at all, the planet would be 59°F (33°C) cooler, like the Moon. As it is, Earth's surface absorbs light energy from the sun and then re-radiates it out as heat energy. This is trapped by greenhouse gases, such as carbon dioxide (CO_2) and methane. Under natural circumstances, these gases are constantly kept in balance between natural "sinks" and "sources," such as CO_2-absorbing rainforests and oceans, and methane-emitting wetlands. However, human activities are now adding gases to the atmosphere at a rate that has overwhelmed the planet's ability to process them through the natural system.

Plants are at the forefront of climate change. Those that can tolerate a broad range of temperatures or are able to disperse their seeds far and wide may not find the new climate regime a problem. But those that are highly adapted to their environment and cannot spread their seeds could become rare or die out altogether. Some scientists have suggested we should create another home for ourselves on Mars in case the changes taking place on Earth jeopardize the future of the human race.

One study of such changes, conducted in 2004 by a collaboration of scientists from around the globe, examined six biodiversity-rich regions covering 20 percent of Earth. They predicted that between 15 and 37 percent of all plant and animal species in those regions could be driven to extinction by the changes likely to

Aloe dichotoma
also known as the quiver tree,
is showing signs that it will not adapt to
a warmer climate.

BELOW *Some scientists believe that humans could one day colonize Mars.*

Turning the red planet green

With Earth's population predicted to be nine billion by 2300 and resources such as rainforests being used unsustainably, some scientists have suggested that humans should colonize another planet. The one they have in mind is Mars, as it has supplies of life-sustaining carbon dioxide, water, and oxygen locked up in its soils. They believe it may be possible to make Mars habitable, a process they call terraforming. This would involve thickening the atmosphere (its pressure is currently one percent of Earth's) and warming up the temperature, which can plummet to -76°F (-60°C). Once the atmosphere was thicker and warmer, the soils could potentially be used to sustain agriculture. Plant hunters and botanists would then be tasked with supplying plants and trees from Earth to change the red planet to a green one.

Earlier flowering

In 1952, former Kew botanist Nigel Hepper started recording the dates on which flowers at Kew opened and he continued noting down his observations for 50 years. When he retired, Kew continued studying 100 of the plants, which included trees, shrubs, bulbs, and herbaceous species. These are now known as the Kew 100. Analysis of the data is revealing some startling findings: Noyes lilac, which once flowered in early summer, is tending to flower in late spring. Meanwhile, snowdrops that opened at the end of February in the 1950s have opened in January since the 1990s. In 2008, Narcissus pseudonarcissus *opened on January 16, seven days earlier than the previous record set in 2007. The 2008 date is 52 days ahead of the average opening date for the 1950s, which was March 9.*

ABOVE *Bulldozers clean up the debris washed onto beaches by heavy rains near Poio, in Galicia, Spain. The year 2006 was particularly hard for the northwestern province, with forest fires affecting large tracts of land and subsequent heavy rains spreading the debris along the coast and destroying the Christmas shellfish harvest.*

take place before 2050. One plant already threatened by climate change is *Aloe dichotoma* (quiver tree) which grows in Namibia and South Africa. Observations from 50 sites show that mortality where the trees grow on slopes is higher at lower elevations, where the temperature is warmer. Mortality rates are also greater in the north of the trees' range, toward the equator, than at the cooler southern margins.

Scientists have begun using phenological records to see how different plants are responding to climate change. Phenology is the science of monitoring the times at which recurring seasonal events take place, such as when trees bud in spring or leaves turn red in the fall. In the largest study so far, conducted in 2006, scientists from 17 countries analyzed 125,000 records and observations made across 21 European countries between 1971 and 2000. From records of 542 plant species and 19 animal species, they concluded that 78 percent of all leafing, flowering, and fruiting is now happening earlier in the year. On average, the European spring is arriving six to eight days earlier than in the past and advancing at 2.5 days per decade.

In the United Kingdom, large trees appear to be responding to elevated temperatures at different rates, with the sycamore, hawthorn, and hornbeam coming into leaf noticeably earlier, and ash and beech exhibiting smaller changes. Over time, this variation will likely alter the competitive advantage of certain species, thus altering the composition of woodlands. For example, the large leaves of the introduced sycamore may come to dominate deciduous woods because they tend to shade out later-leafing native trees. In the southern U.S., plants are coming into leaf later. Scientists believe this is because many need a cold snap to signal when they should begin leafing; the

elevated southern temperatures mean the plants are not getting the message.

Some plants that use cold spells as a signal to break their dormancy are grown commercially; blackcurrants are one example. How these and other food crops respond to changes in climate is of crucial importance to humans. Because many crops have been cultivated for centuries, they have lost their genetic diversity and are therefore more susceptible to climatic changes. "Although there are 30,000 edible plants on the planet, we currently rely on around a dozen," says Professor Stephen Hopper, Director of Kew Gardens (2007-12). One role for the next generation of plant hunters, therefore, will be to seek out new food crops that can tolerate the new climatic regime. What they find out in the world's humid rainforests and arid savannas could shape the future of the entire human race.

BELOW *In the U.K., trees are responding at different rates to the elevated temperatures caused by climate change.*

Index

Numbers in *italics* refer to images